CRITERIA FOR SCIENTIFIC DEVELOPMENT:
PUBLIC POLICY AND NATIONAL GOALS

CRITERIA FOR SCIENTIFIC DEVELOPMENT: PUBLIC POLICY AND NATIONAL GOALS

A Selection of Articles from *Minerva*

Edited by
Edward Shils

THE M.I.T. PRESS

MASSACHUSETTS INSTITUTE OF TECHNOLOGY
CAMBRIDGE, MASSACHUSETTS, AND LONDON, ENGLAND

INTRODUCTION

Edward Shils

Science policy in the form in which is it practiced today is a new thing in human history. This does not mean that earlier societies and epochs had no scientific policies at all. Princely and ecclesiastical patrons of the *ancien régime* frequently attempted to further learning and science by their patronage, and they even did so with quite particular intentions in mind, such as the improvement of navigation or hydraulic engineering. Their intentions toward science were, however, superficial. They were also discontinuous and fragmentary. They had no sense of responsibility for its many-sided development because they had no responsibility for such development. They presupposed the ongoingness of a world of science with a life of its own, which they could influence marginally by prizes for research already done, by rewards for inventions made with the inventor's own resources. Those who reflected on the value of science for improving the material and moral qualities of human life recommended support either for science in general or for particular fields of technology.

They had no concept of science as an internally differentiated and interdependent whole. They were consequently not apprehensive that their "science policy" decisions and actions in one field of science might have a negative or an insufficiently positive bearing on other fields of science. And they did not, apart from interest in practical applications for certain military and economic problems, believe that particular fields of science had strategic significance. Until the French Revolution they did not conceive of scientific actions as being parts of a single social system. Since they did not see scientific activity as forming a whole, they could not think of strategic decisions that might affect more than the specific fields in which the research they wished to support was performed. Science was an already existing phenomenon that could be appreciated, adopted, used, and rewarded. It could not be generated or steered. Regarding science as having a self-generating existence, they did not generally interest themselves in the training of scientists. Higher education and science were not regarded as integrally connected with each other—perhaps because until the nineteenth century, they were not so connnected with each other.

In the past, the actions of the patrons of science and of those scientists who owned whatever resources they needed for their own work certainly influenced the development of science. The influences they exerted were not, however, concerted. There were nearly as many separate decisions as to what should be investigated as there were invesitgators. The concert that existed was the unintended result of the reciprocal responsiveness of approximate equals, none of whom had any authority over the others except for that of intellectual superiority. There were indeed social mechanisms that influenced the growth of science. But these mechanisms did not operate through the authoritative decisions of a corporate body or bodies which took as their main task the guidance of the course of science. Nor was it thought that there should be. That is why science policy in the contemporary sense did not exist.

As long as the choices about what should be studied were widely dispersed over the whole scientific population and their patrons, most of whom possessed the resources to attempt to do whatever they themselves chose, the problem of deciding, over large areas of the scientific scene and even over the totality of science, what was more important and what was less important, did not arise. These conditions no longer obtain. The context of scientific research has changed. The cost of scientific inquiries has increased greatly; each one costs much more than scientific inquiries used to cost, and the total number of inquiries has increased. The present age has at the same time witnessed the institutional concentration of the power to provide the requisite financial resources for almost every important activity, and, accompanying this, there has grown up the belief that it is right and necessary that the allocation of financial resources should be performed in that way.

The increased numbers of scientists and the increased cost of individual investigations have increased the total financial requirements of science. These increases of numbers and costs have practically overwhelmed the capacity of individual scientists to provide for their own research and thus to choose their research problems entirely on their own and in accordance with their own conceptions of what is important, against the background of the received tradition of science. The funds for science have had therefore to come increasingly from the state and from a relatively small number of very wealthy institutions, such as large industrial firms and philanthropic foundations. In relation to the total body of scientists in advanced countries, the number of sources of finances has greatly contracted. In underdeveloped countries, practically all support for research comes from a single source, the state.

This concentration of the capacity for financial provision to a relatively small number of very rich institutions and the corresponding shrinkage of the centers of decision as to the fields of research to be cultivated have coincided with an increased demand for financial resources for many other activities, such as public health, welfare, education, housing, economic growth, defense, transportation, urban renewal, to say nothing of the demands of the vast body of public servants for their own maintenance. Although never as rich as they are now, these institutions are perhaps more conscious than they have ever been of the fundamental fact of scarcity. There is a scarcity of resources to do all the things for which the prospective beneficiaries of expenditure clamor.

Scarcity, in an epoch in which rationality and efficiency have to a much greater extent than heretofore become the criteria for the assessment of policy and performance, imposes the notion of priority. The growing rationality of the technique of budgeting, to say nothing of the Promethean aspiration to plan comprehensively for the future, has imposed a belief that resources should be allocated to diverse uses in accordance with the importance of those uses. Heads of expenditure have to be ranked in accordance with the intrinsic and instrumental value of the activities to be supported by the expenditure. Because science has only a relatively small political constituency, rational arguments rather than political pressure must be invoked to support its claims for a high place in the budget of governments. It, more than most of the activities that governments support, has to be justified by rational arguments about the advantages that flow from it. The fact that the active scientists and scientist-administrators who must provide these arguments are habituated to rational thought has meant that a rational cast of mind confronts the problem of priority. The facts that large expenditures are involved and that among the leading arguments are those who contend that support is justified by the economic advantages of scientific research have brought economists into the discussion.

In a convergent movement with this has grown the *systemic* mode of thought. There is a heightened sensitivity nowadays to the interdependence of events. Economic theory does not by any means enjoy a monopoly in the domination of governmental policies or public opinion, but the idea of an optimal allocation of scarce resources among alternative uses—of an allocation that strives to attain equality among the marginal products of the alternative uses of scarce resources—has taken root. Science too has come to be viewed in the light of this "optimum." There is now a belief that a "right order" of expenditure is conceivable and attainable, and this applies to science as well as to other fields of human activity. The systemic mode of thought postulates scarcity; it also postulates determinate patterns of interdependence among the variables, the scarcity of which imposes "economy." The acknowledgment of a linear time sequence of processes imposes inescapable tasks on policies that aspire to rationality.

The conception of the comprehensive planning of society, and of the economy in particular, has added another fundamental ingredient to the combination of forces that has fostered the idea of a comprehensive and rational science policy. The notion that a whole society could be planned deliberately in a way that would shape it for a long time to come presupposed not only a pervasive knowledge of the present state of society but the ability to foresee the subsequent behavior of its component parts. The growing prestige of the idea of planning and the growing courage of those who would predict the future have strengthened the conviction that a rational science policy is necessary and possible.

The notion of a right order of scientific activities—of an optimal allocation of financial and manpower resources for science—would still not have become the object of science policy if it were not also believed that this optimum could be realized by deliberate central decision. Various spheres of culture have been subjected to efforts of central control in the past, and some continue to be. Literature and art are subjected to censorship; but in most countries the censorship is fairly marginal, and it is, in any case, negative. Propaganda for particular forms of literature and art and the offer of publication and distribution facilities only to works of a certain type and outlook represent a more positive effort to guide the content of literary and artistic output. Likewise, the religious sphere has often known intolerance and suppression of certain institutional manifestations of religious belief, and there have also been strenuous efforts to coerce the populace to accept the doctrines of one particular religion. But neither the control over literature and art nor the control over religion has been intended to foster the *development* of those spheres of culture. They were concerned rather to propagate or suppress *already existing* beliefs and practices. They were not intended to promote creativity. The intention to promote creativity—the nurturing of tendencies hitherto unrealized, the cultivation of the previously unknown—is the unique feature of contemporary science policy, in comparison with the "science policies" of the past and in comparison with policies in other fields of governmental action.

The present-day encouragement of the performance of previously unperformed acts appreciates that science has a life of its own, which, however much it can be affected from the outside, remains the essence of scientific activity. Science is acknowledged to possess an irreducible autonomy that cannot be replaced. This is a process internal to science; if it is suspended, then science ceases to operate. All science policy can do is to influence the working of this autonomous system through decisions that grant (or withhold) financial resources, provide an *appropriate* administrative context, supply manpower, or set certain tasks. Once these are done, the autonomously systemic properties of scientific activity must be allowed to manifest themselves. This view of science and of the potentialities of science policy has now come to be accepted very widely, even by those who espouse a far-reaching and comprehensive planning of science.

These developments have come about piecemeal and in an uncoordinated manner. The science policies that have grown up as their result are characterized by a similar incoherence. Nonetheless, and perhaps even because of the very incoherence of present-day science policies, there is a genuine aspiration to make science policy more rational.

The science policy at which the present discussion of science policy aims is the deliberate effort to influence the direction and rate of the development of scientific knowledge through the application of financial resources, administrative devices, and education and training in so far as these are affected by political authority. The accomplishments of individual scientists constitute scientific development, and the exemplariness and persuasiveness of the performance of the greatest among them are certainly major determinants of the direction and rate of development of scientific knowledge. But the exercise of influence through the discovery and promulgation of new scientific knowledge is not the kind of influence that we mean when we speak nowadays of science policy. In science policy, the decision to influence and the action that influences are decisions outside the constitution of scientific activity itself. A decision of political authority to allow complete autonomy to every sector of the scientific community—as might be implied by Professor Polanyi's conception of the "republic of science"—would be an act of scientific policy. But the factual existence of such an autonomy, which resulted from the traditional dissociation of political authority from the scientific sphere, would be a consequence of science policy. A rational and comprehensive science policy involves the intention to influence scientific development through authoritative decisions,[1] which choose particular problems or whole fields

---

[1] One may speak of the science policy of a particular industrial firm as well as of the science policy or policies of a government. Yet one would not be likely to speak of the scientific policy of a particular scientist with regard to his own scientific activity. The concept of scientific policy refers to the macrosocial system of scientific activity and to decisions made outside the system of scientific activity as such.

of inquiry, or which fix the institutional setting within which scientists themselves choose the problems on which they will work or the fields which they wish to cultivate. The decisions might refer to narrow sectors, to broad sectors, or conceivably to science as a whole.

The justification for science policy is that the decisions it produces will be rational decisions taken in the light of the ends to be attained—the ends being the development of science and the application of scientific knowledge, that is, knowledge gained through systematic research for practical ends. The numerous particular decisions could conceivably constitute a more or less rational pattern or system of decisions.

At present, every country which has a substantial amount of scientific activity, even many of those which have very little, has something like an empirical science policy or, perhaps it would be more accurate to say, science policies. It is not, however, unjust to say that none has a rational and comprehensive science policy. None has a science policy in the sense which the papers which form this book seek to realize. What exists is a large amount of influence exercised by governmental and private bodies that are not themselves constituted by scientists; the decisions taken by these bodies are uncoordinated with one another, and most of them are directed not to the scientific system as a whole but to particular parts of science without much regard to their relation to other parts of science or to the educational system. Where attention is paid to problems of coordination, criteria of judgment and assessments of magnitude are extremely vague and are applied in a very inconstant way. Decisions are made on the basis of political considerations to satisfy domestic pressure groups, to build personal and departmental "empires," and to compete for international prestige, as well as on the basis of relatively well considered beliefs about the potential contribution of research to the realization of ends such as the improvement of industrial and agricultural and physical and mental health, and military technology. The increased frequency of decisions that affect science and that are made outside the scientific system itself has accentuated the demand and aspiratións of scientists and science-administrators for better ways of making science policy, for a science policy that embodies some explicitly articulated and rational principles.

The idea of "planning of science" represented an attempt to rationalize science policy. The idea of the "republic of science" was likewise an attempt to introduce rationality into science policy. The most important thing that has emerged from the discussion in *Minerva* and elsewhere in recent years has been the acceptance of two independent and incommensurable criteria of scientific choice: scientific value and practical value. The distinction is not a new one. It is roughly parallel to the distinction between pure science and applied science. It has similar parallels in the distinction between the immanent dynamics of scientific growth and the determination of the direction of scientific development by economic tasks. political motives, and so on. Another related distinction, formulated in the idiom of economic analysis but entailing similar substantive differences, is that between scientific research as a "consumer's good" and as an "investment." These distinctions correspond to the divergent outlooks which praise, respectively, the autonomy of science and the comprehensive direction of science toward technological applications.

The "planning of science" has turned out to have been a cover for varying combinations of arbitrary political imposition, individual and institutional "empire building," *laissez faire,* and sheer disorder. Complete *laissez faire* is patently impracticable in situations where decisions *must* inevitably be made concerning the allocation of huge sums of money for conflicting and competing scientific projects, where there is a single or very small number of sources of funds, and where there is an urgent and evident need for research directed toward the improvement of welfare in various fields of medical, industrial, and agricultural, to say nothing of military technology.

While it is clear that there are as yet no satisfactory principles of science policy capable of realistic and thoroughgoing application to the multifarious activities of science, progress is being made in their discernment and clarification. The progress is partly negative: it consists of the renunciation of the extreme positions which once claimed universal validity.

The abandonment of extreme positions is, however, only a necessary first step in any realistic approach to science policy. Once this step has been taken, the complexity and multiplicity of the situations about which decisions must be made are laid open to freer consideration. It is now seen that there is neither a single goal nor a unitary set of goals toward which science as a whole can be planned, that there is no single institutional arrangement that is equally appropriate to the development of all its parts, that there is no inevitable harmony between the development of all branches of science and every other social, economic, and political need, and that scientific development as such does not automatically and inevitably improve the welfare of mankind. It is now seen that scientific policy has never been planned in any way satisfactory to scientists and to those who hope that their particular ends, economic, political, social, and cultural, would be aided by scientific growth. In fact, it has never been planned at all as planning is understood by its proponents. Like every other human activity, scientific activity exists in the context of a scarcity of funds and of personnel, and the ends that it might serve are in competition with each other.

The discussions of the criteria of scientific choice which are contained in this book represent the beginnings of a movement toward the more rational science policy that is felt to be so necessary. The theory of science policy is still very rudimentary. It is still very general in its reference to the whole range of scientific activities and those educational, technological, political, military, and administrative activities in which science is involved as cause and effect.

Nonetheless it is under way. The present collection of essays drawn from the first five years of *Minerva* attests to the efforts that have been made in recent years by science administrators and scientists, economists and philosophers, to bring the analysis of the problems of scientific choice to the point where it can begin to be useful. In its present stage at least, the theory of science policy offers no recipes or directives that can be confidently applied to particular decisions. It does, however, offer a clarification of some of the elements that are involved in decisions about expenditures on science. As such its promulgators may legitimately claim to provide improved general canons of judgment and guiding principles to legislators, administrators, advisers, scientists, and citizens who are required to decide what should be supported, to what extent, and in what manner.

For about thirty years, the conflict between the proponents of pure science and the proponents of applied science, between the liberals and the planners, has bedeviled the discussion of scientific policy. The most recent discussions, as expressed in the papers contained in this collection, have gradually ameliorated the tension between these two criteria and the policies of scientific development that were associated with them. As Dr. A. M. Weinberg shows in his papers, scientific choice requires the application of a combination of diverse criteria. The criteria of scientific merit, technological merit, and social merit might be contradictory to one another; a given research scheme might be high in scientific merit and low in the other two. The fact that these criteria are sometimes and perhaps even often incompatible does not mean that they are not equally valid.

Their validity does not render them capable of easy application. Even though they represent a considerable progress in the discussion, they are nonetheless vague and undifferentiated. It is, moreover, difficult to estimate probabilities of scientific or technological fruitfulness; it is at present impossible to assess the value of a scientific discovery in one field as against one in another field, even within pure science. And how is one to assess the value of one plausibly predictable scientific outcome against the value of an equally plausibly predictable increment to human welfare arising from scientific research?

All this being granted, it seems to me undeniable that the essays in this collection which deal with scientific choice have broken new ground. They have made distinctions that have to be made and made possible more reasonable and wiser judgments. They are not yet a code of rational scientific choice. It might be that such a code of rational scientific choice cannot in the nature of things be attained. Yet it is certain that a closer approximation to that goal is possible. Even if a fully rational policy is unattainable, a more rational discussion of the alternatives of policy is attainable. These essays should be regarded as contributions to the movement in that direction.

The elucidation of the criteria of rational scientific choice does not exhaust the tasks with which such principles would have to cope. Organizational or administrative problems arise at once from every suggestion of a principle of choice. If the principle is to be wisely applied by decision makers, the latter will need qualified advisers. How should advisers be chosen and employed, what should be their powers, and how should the flow of their advice be organized? What should be the terms of reference of these advisers, under what conditions should they serve their own scientific intertst, and under what conditions, and how, should this latter interest be guarded against?

If research is to be applied for industrial, agricultural, or welfare purposes, where should the center of gravity of the decision-making machinery be located? Should it be located in the industrial, agricultural, or welfare departments or institutions that will use it? Or should it be kept separate from the "operating agencies," and, if separate, then in just what way? Regardless of the location of decision, how are laboratories and research institutions best organized to enhance creativity and efficiency?

Similarly, whatever the criteria of scientific choice to be applied, what are the best means of ensuring the flow of the right numbers of properly qualified and motivated research workers to those research projects which are chosen? How, in what sense, and to what extent can the future demand for research workers, science teachers, and so on, be predicted, and to what extent and how can their supply be planned?

For underdeveloped countries, many or most of the problems of scientific policy are the same as those of the advanced countries. There is one very important exception. This is the establishment of a scientific tradition, that is, the establishment of beliefs and orientations that heighten and maintain sensibilities and motivations and that prompt the selection of important and appropriate problems for investigation and suggest the approach toward them in ways that permit their fruitful solution. Countries in which science is well established may take this for granted. All they need do is to see that there is a flow of young students into fields needing investigation and into institutions in which work is being done in those fields. The students will then become assimilated into the scientific tradition. Such conditions do not obtain in underdeveloped countries, and it is an obvious task of science policy in these countries to make the arrangements that will foster the establishment of such a tradition.

Thus it may be seen that the problem of scientific choice is only one facet, albeit a very crucial one, of any approximately rational science policy. Its further development stands in need of research and analysis in many ancillary fields, such as the newly developing subject of the sociology of science, which deals with the social structure of research institutions, and the social conditions of the growth of science, the new "science of science," which deals with rates and magnitudes of scientific growth, the psychology of science, which deals with the processes and conditions of creativity, the political science of science, which deals with the relations between politicians, administrators, and scientists, and that nameless field that deals with the optimal conditions for the translation of the results of research into economic growth. We need very much more exact factual knowledge about the community of science and its relations with the rest of society.

When I began *Minerva* in 1962, I sketched a wide-ranging agenda that embraced every aspect of the social, economic, moral, political, and administrative relations of scientific research and higher education: the influence of the increased demands of governments on science and learning, the influence of the increased munificence of governments on science and learning, the consequences of the increased demands of scientists for support for their boundless curiosity, and the increased demands of society for higher levels of welfare, which require continuous investment in research. Improved understanding of the relations between government and systematic and disciplined inquiry in science and scholarship was taken as the subject matter of *Minerva*. "By the improvement of understanding," I wrote on the opening page of the first issue, "it [*Minerva*] hopes to make scientific and academic policy more reasonable and realistic." I believe that the essays that follow show that we have not stood still.

The owl of *Minerva* has not waited for the shades of night to fall before taking flight. On the contrary, it has made itself into a carrier of light to illuminate a subject, the obscure complexity of which corresponds to its importance for our intellectual and material well-being.

CONTENTS

THE REPUBLIC OF SCIENCE

Its Political and Economic Theory

MICHAEL POLANYI

MY title is intended to suggest that the community of scientists is organised in a way which resembles certain features of a body politic and works according to economic principles similar to those by which the production of material goods is regulated. Much of what I will have to say will be common knowledge among scientists, but I believe that it will recast the subject from a novel point of view which can both profit from and have a lesson for political and economic theory. For in the free cooperation of independent scientists we shall find a highly simplified model of a free society, which presents in isolation certain basic features of it that are more difficult to identify within the comprehensive functions of a national body.

The first thing to make clear is that scientists, freely making their own choice of problems and pursuing them in the light of their own personal judgment are in fact cooperating as members of a closely knit organisation. The point can be settled by considering the opposite case where individuals are engaged in a joint task without being in any way coordinated. A group of women shelling peas work at the same task, but their individual efforts are not coordinated. The same is true of a team of chess players. This is shown by the fact that the total amount of peas shelled and the total number of games won will not be affected if the members of the group are isolated from each other. Consider by contrast the effect which a complete isolation of scientists would have on the progress of science. Each scientist would go on for a while developing problems derived from the information initially available to all. But these problems would soon be exhausted, and in the absence of further information about the results achieved by others, new problems of any value would cease to arise and scientific progress would come to a standstill.

This shows that the activities of scientists are in fact coordinated, and it also reveals the principle of their coordination. This consists in the adjustment of the efforts of each to the hitherto achieved results of the others. We may call this a coordination by mutual adjustment of independent initiatives —of initiatives which are coordinated because each takes into account all the other initiatives operating within the same system.

This article from *Minerva*, I, 1 (Autumn, 1962), pp. 54–73.

1

WHEN put in these abstract terms the principle of spontaneous coordination of independent initiatives may sound obscure. So let me illustrate it by a simple example. Imagine that we are given the pieces of a very large jig-saw puzzle, and suppose that for some reason it is important that our giant puzzle be put together in the shortest possible time. We would naturally try to speed this up by engaging a number of helpers; the question is in what manner these could be best employed. Suppose we share out the pieces of the jig-saw puzzle equally among the helpers and let each of them work on his lot separately. It is easy to see that this method, which would be quite appropriate to a number of women shelling peas, would be totally ineffectual in this case, since few of the pieces allocated to one particular assistant would be found to fit together. We could do a little better by providing duplicates of all the pieces to each helper separately, and eventually somehow bring together their several results. But even by this method the team would not much surpass the performance of a single individual at his best. The only way the assistants can effectively cooperate and surpass by far what any single one of them could do, is to let them work on putting the puzzle together in sight of the others, so that every time a piece of it is fitted in by one helper, all the others will immediately watch out for the next step that becomes possible in consequence. Under this system, each helper will act on his own initiative, by responding to the latest achievements of the others, and the completion of their joint task will be greatly accelerated. We have here in a nutshell the way in which a series of independent initiatives are organised to a joint achievement by mutually adjusting themselves at every successive stage to the situation created by all the others who are acting likewise.

Such self-coordination of independent initiatives leads to a joint result which is unpremeditated by any of those who bring it about. Their coordination is guided as by ' an invisible hand ' towards the joint discovery of a hidden system of things. Since its end-result is unknown, this kind of cooperation can only advance stepwise, and the total performance will be the best possible if each consecutive step is decided upon by the person most competent to do so. We may imagine this condition to be fulfilled for the fitting together of a jig-saw puzzle if each helper watches out for any new opportunities arising along a particular section of the hitherto completed patch of the puzzle, and also keeps an eye on a particular lot cf pieces, so as to fit them in wherever a chance presents itself. The effectiveness of a group of helpers will then exceed that of any isolated member, to the extent to which some member of the group will always discover a new chance for adding a piece to the puzzle more quickly than any one isolated person could have done by himself.

Any attempt to organise the group of helpers under a single authority would eliminate their independent initiatives and thus reduce their joint effectiveness to that of the single person directing them from the centre. It would, in effect, paralyse their cooperation.

Essentially the same is true for the advancement of science by independent initiatives adjusting themselves consecutively to the results achieved by all the others. So long as each scientist keeps making the best contribution of which he is capable, and on which no one could improve (except by abandoning the problem of his own choice and thus causing an overall loss to the advancement of science), we may affirm that the pursuit of science by independent self-coordinated initiatives assures the most efficient possible organisation of scientific progress. And we may add, again, that any authority which would undertake to direct the work of the scientist centrally would bring the progress of science virtually to a standstill.

WHAT I have said here about the highest possible coordination of individual scientific efforts by a process of self-coordination may recall the self-coordination achieved by producers and consumers operating in a market. It was, indeed, with this in mind that I spoke of ' the invisible hand ' guiding the coordination of independent initiatives to a maximum advancement of science, just as Adam Smith invoked ' the invisible hand ' to describe the achievement of greatest joint material satisfaction when independent producers and consumers are guided by the prices of goods in a market. I am suggesting, in fact, that the coordinating functions of the market are but a special case of coordination by mutual adjustment. In the case of science, adjustment takes place by taking note of the published results of other scientists; while in the case of the market, mutual adjustment is mediated by a system of prices broadcasting current exchange relations, which make supply meet demand.

But the system of prices ruling the market not only transmits information in the light of which economic agents can mutually adjust their actions; it also provides them with an incentive to exercise economy in terms of money. We shall see that, by contrast, the scientist responding directly to the intellectual situation created by the published results of other scientists is motivated by current professional standards.

Yet in a wider sense of the term, the decisions of a scientist choosing a problem and pursuing it to the exclusion of other possible avenues of inquiry may be said to have an economic character. For his decisions are designed to produce the highest possible result by the use of a limited stock of intellectual and material resources. The scientist fulfils this purpose by choosing a problem that is neither too hard nor too easy for him. For to

3

apply himself to a problem that does not tax his faculties to the full is to waste some of his faculties; while to attack a problem that is too hard for him would waste his faculties altogether. The psychologist K. Lewin has observed that one's person never becomes fully involved either in a problem that is much too hard, nor in one that is much too easy. The line the scientist must choose turns out, therefore, to be that of greatest ego-involvement; it is the line of greatest excitement, sustaining the most intense attention and effort of thought. The choice will be conditioned to some extent by the resources available to the scientist in terms of materials and assistants, but he will be ill-advised to choose his problem with a view to guaranteeing that none of these resources be wasted. He should not hesitate to incur such a loss, if it leads him to deeper and more important problems.

THIS is where professional standards enter into the scientist's motivation. He assesses the depth of a problem and the importance of its prospective solution primarily by the standards of scientific merit accepted by the scientific community—though his own work may demand these standards to be modified. Scientific merit depends on a number of criteria which I shall enumerate here under three headings. These criteria are not altogether independent of each other, but I cannot analyse here their mutual relationship.

(1) The first criterion that a contribution to science must fulfil in order to be accepted is a sufficient degree of plausibility. Scientific publications are continuously beset by cranks, frauds and bunglers whose contributions must be rejected if journals are not to be swamped by them. This censorship will not only eliminate obvious absurdities but must often refuse publication merely because the conclusions of a paper appear to be unsound in the light of current scientific knowledge. It is indeed difficult even to start an experimental inquiry if its problem is considered scientifically unsound. Few laboratories would accept today a student of extra-sensory perception, and even a project for testing once more the hereditary transmission of acquired characters would be severely discouraged from the start. Besides, even when all these obstacles have been overcome, and a paper has come out signed by an author of high distinction in science, it may be totally disregarded, simply for the reason that its results conflict sharply with the current scientific opinion about the nature of things.

I shall illustrate this by an example which I have used elsewhere (*The Logic of Liberty*, London and Chicago, 1951, p. 12). A series of simple experiments were published in June 1947 in the *Proceedings of the Royal Society* by Lord Rayleigh—a distinguished Fellow of the Society—purporting to show that hydrogen atoms striking a metal wire transmit to it energies up to a hundred electron volts. This, if true, would have been far more revolutionary than the discovery of atomic fission by Otto Hahn. Yet, when I asked physicists what they thought about it, they only shrugged their shoulders. They could not find fault with the experiment yet not one believed in its results, nor thought it worth while to repeat it. They just ignored it. A possible explanation of Lord Rayleigh's experiments is given in my *Personal Knowledge* (1958) p. 276. It appears that the physicists missed nothing by disregarding these findings.

(2) The second criterion by which the merit of a contribution is assessed, may be described as its scientific value, a value that is composed of the following three coefficients: (a) its accuracy, (b) its systematic importance, (c) the intrinsic interest of its subject-matter. You can see these three gradings entering jointly into the value of a paper in physics compared with one in biology. The inanimate things studied by physics are much less interesting than the living beings which are the subject of biology. But physics makes up by its great accuracy and wide theoretical scope for the dullness of its subject, while biology compensates for its lack of accuracy and theoretical beauty by its exciting matter.

(3) A contribution of sufficient plausibility and of a given scientific value may yet vary in respect of its originality; this is the third criterion of scientific merit. The originality of technical inventions is assessed, for the purpose of claiming a patent, in terms of the degree of surprise which the invention would cause among those familiar with the art. Similarly, the originality of a discovery is assessed by the degree of surprise which its communication should arouse among scientists. The unexpectedness of a discovery will overlap with its systematic importance, yet the surprise caused by a discovery, which causes us to admire its daring and ingenuity, is something different from this. It pertains to the act of producing the discovery. There are discoveries of the highest daring and ingenuity, as for example the discovery of Neptune, which have no great systematic importance.

BOTH the criteria of plausibility and of scientific value tend to enforce conformity, while the value attached to originality encourages dissent. This internal tension is essential in guiding and motivating scientific work. The professional standards of science must impose a framework of discipline and at the same time encourage rebellion against it. They must demand that, in order to be taken seriously, an investigation should largely conform to the currently predominant beliefs about the nature of things, while allowing that in order to be original it may to some extent go against these. Thus, the authority of scientific opinion enforces the teachings of

5

science in general, for the very purpose of fostering their subversion in particular points.

This dual function of professional standards in science is but the logical outcome of the belief that scientific truth is an aspect of reality and that the orthodoxy of science is taught as a guide that should enable the novice eventually to make his own contacts with this reality. The authority of scientific standards is thus exercised for the very purpose of providing those guided by it with independent grounds for opposing it. The capacity to renew itself by evoking and assimilating opposition to itself appears to be logically inherent in the sources of the authority wielded by scientific orthodoxy.

But who is it, exactly, who exercises the authority of this orthodoxy? I have mentioned scientific opinion as its agent. But this raises a serious problem. No single scientist has a sound understanding of more than a tiny fraction of the total domain of science. How can an aggregate of such specialists possibly form a joint opinion? How can they possibly exercise jointly the delicate function of imposing a current scientific view about the nature of things, and the current scientific valuation of proposed contributions, even while encouraging an originality which would modify this orthodoxy? In seeking the answer to this question we shall discover yet another organisational principle that is essential for the control of a multitude of independent scientific initiatives. This principle is based on the fact that, while scientists can admittedly exercise competent judgment only over a small part of science, they can usually judge an area adjoining their own special studies that is broad enough to include some fields on which other scientists have specialised. We thus have a considerable degree of overlapping between the areas over which a scientist can exercise a sound critical judgment. And, of course, each scientist who is a member of a group of overlapping competences will also be a member of other groups of the same kind, so that the whole of science will be covered by chains and networks of overlapping neighbourhoods. Each link in these chains and networks will establish agreement between the valuations made by scientists overlooking the same overlapping fields, and so, from one overlapping neighbourhood to the other, agreement will be established on the valuation of scientific merit throughout all the domains of science. Indeed, through these overlapping neighbourhoods uniform standards of scientific merit will prevail over the entire range of science, all the way from astronomy to medicine. This network is the seat of scientific opinion. Scientific opinion is an opinion not held by any single human mind, but one which, split into thousands of fragments, is held by a multitude of individuals, each of whom

6

endorses the other's opinion at second hand, by relying on the consensual chains which link him to all the others through a sequence of overlapping neighbourhoods.

ADMITTEDLY, scientific authority is not distributed evenly throughout the body of scientists; some distinguished members of the profession predominate over others of a more junior standing. But the authority of scientific opinion remains essentially mutual; it is established *between* scientists, not above them. Scientists exercise their authority over each other. Admittedly, the body of scientists, as a whole, does uphold the authority of science over the lay public. It controls thereby also the process by which young men are trained to become members of the scientific profession. But once the novice has reached the grade of an independent scientist, there is no longer any superior above him. His submission to scientific opinion is entailed now in his joining a chain of mutual appreciations, within which he is called upon to bear his equal share of responsibility for the authority to which he submits.

Let me make it clear, even without going into detail, how great and varied are the powers exercised by this authority. Appointments to positions in universities and elsewhere, which offer opportunity for independent research, are filled in accordance with the appreciation of candidates by scientific opinion. Referees reporting on papers submitted to journals are charged with keeping out contributions which current scientific opinion condemns as unsound; and scientific opinion is in control, once more, over the issue of textbooks, as it can make or mar their influence through reviews in scientific journals. Representatives of scientific opinion will pounce upon newspaper articles or other popular literature which would venture to spread views contrary to scientific opinion. The teaching of science in schools is controlled likewise. And, indeed, the whole outlook of man on the universe is conditioned by an implicit recognition of the authority of scientific opinion.

I have mentioned earlier that the uniformity of scientific standards throughout science makes possible the comparison between the value of discoveries in fields as different as astronomy and medicine. This possibility is of great value for the rational distribution of efforts and material resources throughout the various branches of science. If the minimum merit by which a contribution would be qualified for acceptance by journals were much lower in one branch of science than in another, this would clearly cause too much effort to be spent on the former branch as compared with the latter. Such is in fact the principle which underlies the rational distribution of grants for the pursuit of research. Subsidies should be curtailed in areas

7

where their yields in terms of scientific merit tend to be low, and should be channelled instead to the growing points of science, where increased financial means may be expected to produce a work of higher scientific value. It does not matter for this purpose whether the money comes from a public authority or from private sources, nor whether it is disbursed by a few sources or a large number of benefactors. So long as each allocation follows the guidance of scientific opinion, by giving preference to the most promising scientists and subjects, the distribution of grants will automatically yield the maximum advantage for the advancement of science as a whole. It will do so, at any rate, to the extent to which scientific opinion offers the best possible appreciation of scientific merit and of the prospects for the further development of scientific talent.

For scientific opinion may, of course, sometimes be mistaken, and as a result unorthodox work of high originality and merit may be discouraged or altogether suppressed for a time. But these risks have to be taken. Only the discipline imposed by an effective scientific opinion can prevent the adulteration of science by cranks and dabblers. In parts of the world where no sound and authoritative scientific opinion is established research stagnates for lack of stimulus, while unsound reputations grow up based on commonplace achievements or mere empty boasts. Politics and business play havoc with appointments and the granting of subsidies for research; journals are made unreadable by including much trash.

Moreover, only a strong and united scientific opinion imposing the intrinsic value of scientific progress on society at large can elicit the support of scientific inquiry by the general public. Only by securing popular respect for its own authority can scientific opinion safeguard the complete independence of mature scientists and the unhindered publicity of their results, which jointly assure the spontaneous coordination of scientific efforts throughout the world. These are the principles of organisation under which the unprecedented advancement of science has been achieved in the twentieth century. Though it is easy to find flaws in their operation, they yet remain the only principles by which this vast domain of collective creativity can be effectively promoted and coordinated.

DURING the last 20 to 30 years, there have been many suggestions and pressures towards guiding the progress of scientific inquiry in the direction of public welfare. I shall speak mainly of those I have witnessed in England. In August 1938 the British Association for the Advancement of Science founded a new division for the social and international relations of science, which was largely motivated by the desire to offer deliberate social guidance to the progress of science. This programme was given more extreme expression by the Association of Scientific Workers in Britain. In January 1943 the Association filled a large hall in London with a meeting attended by many of the most distinguished scientists of the country, and it decided—in the words officially summing up the conference—that research would no longer be conducted for itself as an end in itself. Reports from Soviet Russia describing the successful conduct of scientific research, according to plans laid down by the Academy of Science, with a view to supporting the economic Five-Year Plans, encouraged this resolution.

I appreciate the generous sentiments which actuate the aspiration of guiding the progress of science into socially beneficent channels, but I hold its aim to be impossible and nonsensical.

An example will show what I mean by this impossibility. In January 1945 Lord Russell and I were together on the BBC Brains Trust. We were asked about the possible technical uses of Einstein's theory of relativity, and neither of us could think of any. This was 40 years after the publication of the theory and 50 years after the inception by Einstein of the work which led to its discovery. It was 58 years after the Michelson-Morley experiment. But, actually, the technical application of relativity, which neither Russell nor I could think of, was to be revealed within a few months by the explosion of the first atomic bomb. For the energy of the explosion was released at the expense of mass in accordance with the relativistic equation $e = mc^2$, an equation which was soon to be found splashed over the cover of *Time* magazine, as a token of its supreme practical importance.

Perhaps Russell and I should have done better in foreseeing these applications of relativity in January 1945, but it is obvious that Einstein could not possibly take these future consequences into account when he started on the problem which led to the discovery of relativity at the turn of the century. For one thing, another dozen or more major discoveries had yet to be made before relativity could be combined with them to yield the technical process which opened the atomic age.

Any attempt at guiding scientific research towards a purpose other than its own is an attempt to deflect it from the advancement of science. Emergencies may arise in which all scientists willingly apply their gifts to tasks of public interest. It is conceivable that we may come to abhor the progress of science, and stop all scientific research or at least whole branches of it, as the Soviets stopped research in genetics for 25 years. You can kill or mutilate the advance of science, you cannot shape it. For it can advance only by essentially unpredictable steps, pursuing problems of its own, and the practical benefits of these advances will be incidental and hence doubly unpredictable.

In saying this, I have *not* forgotten, but merely set aside, the vast amount of scientific work currently conducted in industrial and governmental laboratories [1] In describing here the autonomous growth of science, I have taken the relation of science to technology fully into account.

BUT even those who accept the autonomy of scientific progress may feel irked by allowing such an important process to go on without trying to control the coordination of its fragmentary initiatives. The period of high aspirations following the last war produced an event to illustrate the impracticability of this more limited task.

The incident originated in the University Grants Committee, which sent a memorandum to the Royal Society in the summer of 1945. The document, signed by Sir Charles Darwin, requested the aid of the Royal Society to secure ' The Balanced Development of Science in the United Kingdom '; this was its title.

The proposal excluded undergraduate studies and aimed at the higher subjects that are taught through the pursuit of research. Its main concern was with the lack of coordination between universities in taking up ' rare ' subjects, ' which call for expert study at only a few places, or in some cases perhaps only one '. This was linked with the apprehension that appointments are filled according to the dictates of fashion, as a result of which some subjects of greater importance are being pursued with less vigour than others of lesser importance. It proposed that a coordinating machinery should be set up for levelling out these gaps and redundancies. The Royal Society was asked to compile, through its Sectional Committees covering the main divisions of science, lists of subjects deserving preference in order to fill gaps. Such surveys were to be renewed in the future to guide the University Grants Committee in maintaining balanced proportions of scientific effort throughout all fields of inquiry.

Sir Charles Darwin's proposal was circulated by the Secretaries of the Royal Society to the members of the Sectional Committees, along with a report of previous discussions of his proposals by the Council and other groups of Fellows. The report acknowledged that the coordination of the pursuit of higher studies in the universities was defective (' haphazard ') and endorsed the project for periodic, most likely annual, surveys of gaps and redundancies by the Royal Society. The members of the Sectional Committees were asked to prepare, for consideration by a forthcoming meeting of the Committees, lists of subjects suffering from neglect.

Faced with this request which I considered, at the best, pointless, I wrote to the Physical Secretary (the late Sir Alfred Egerton) to express my doubts.

---

[1] I have analysed the relation between academic and industrial science quite recently and in some detail (J. Inst. Met. *89* (1961) 401.)

I argued that the present practice of filling vacant chairs by the most eminent candidate that the university can attract was the best safeguard for rational distribution of efforts over rival lines of scientific research. As an example (which should appeal to Sir Charles Darwin as a physicist) I recalled the successive appointments to the chair of physics in Manchester during the past thirty years. Manchester had elected to this chair Schuster, Rutherford, W. L. Bragg and Blackett, in this sequence, each of whom represented at the time a ' rare ' section of physics: spectroscopy, radio-activity, X-ray crystallography, and cosmic-rays, respectively. I affirmed that Manchester had acted rightly and that they would have been ill-advised to pay attention to the claims of subjects which had not produced at the time men of comparable ability. For the principal criterion for offering increased opportunities to a new subject was the rise of a growing number of distinguished scientists in that subject and the falling off of creative initiative in other subjects, indicating that resources should be withdrawn from them. While admitting that on certain occasions it may be necessary to depart from this policy, I urged that it should be recognised as the essential agency for maintaining a balanced development of scientific research.

Sir Alfred Egerton's response was sympathetic, and, through him, my views were brought to the notice of the members of Sectional Committees. Yet the Committees met, and I duly took part in compiling a list of ' neglected subjects ' in chemistry. The result, however, appeared so vague and trivial (as I will illustrate by an example in a moment) that I wrote to the Chairman of the Chemistry Committee that I would not support the Committee's recommendations if they should be submitted to the Senate of my university.

However, my worries were to prove unnecessary. Already the view was spreading among the Chairmen of the Sectional Committees ' that a satis-factory condition in each science would come about naturally, provided that each university always chose the most distinguished leaders for its post, irrespective of his specialisation '. While others still expressed the fear that this would make for an excessive pursuit of fashionable subjects, the upshot was, at the best, inconclusive. Darwin himself had, in fact, already declared the reports of the Sectional Committees ' rather disappointing '.

The whole action was brought to a close, one year after it had started, with a circular letter to the Vice-Chancellors of the British universities signed by Sir Alfred Egerton, as secretary, on behalf of the Council of the Royal Society, a copy being sent to the University Grants Committee. The circular included copies of the reports received from the Sectional Committees and endorsed these in general. But in the body of the letter only a small number

11

of these recommendations were specified as being of special importance. This list contained seven recommendations for the establishment of new schools of research, but said nothing about the way these new schools should be coordinated with existing activities all over the United Kingdom. The impact of this document on the universities seems to have been negligible. The Chemistry Committee's recommendation for the establishment of ' a strong school of analytic chemistry ', which should have concerned me as Professor of Physical Chemistry, was never even brought to my notice in Manchester.

I HAVE not recorded this incident in order to expose its error. It is an important historical event. Most major principles of physics are founded on the recognition of an impossibility, and no body of scientists was better qualified than the Royal Society to demonstrate that a central authority cannot effectively improve on the spontaneous emergence of growing points in science. It has proved that little more can, or need, be done towards the advancement of science, than to assist spontaneous movements towards new fields of distinguished discovery, at the expense of fields that have become exhausted. Though special considerations may deviate from it, this procedure must be acknowledged as the major principle for maintaining a balanced development of scientific research.

(Here is the point at which this analysis of the principles by which funds are to be distributed between different branches of science may have a lesson for economic theory. It suggests a way in which resources can be rationally distributed between *any* rival purposes that cannot be valued in terms of money. All cases of public expenditure serving purely collective interests are of this kind. A comparison of such values by a network of overlapping competences may offer a possibility for a true collective assessment of the relative claims of thousands of government departments of which no single person can know well more than a tiny fraction.)

But let me recall yet another striking incident of the post-war period which bears on these principles. I have said that the distribution of subsidies to pure science should not depend on the sources of money, whether they are public or private. This will hold to a considerable extent also for subsidies given to universities as a whole. But after the war, when in England the cost of expanding universities was largely taken over by the state, it was felt that this must be repaid by a more direct support for the national interest. This thought was expressed in July 1946 by the Committee of Vice-Chancellors in a memorandum sent out to all universities, which Sir Ernest Simon (as he then was) as Chairman of the Council of

Manchester University, declared to be of 'almost revolutionary' importance. I shall quote a few extracts:

> The universities entirely accept the view that the Government has not only the right, but the duty, to satisfy itself that every field of study which in the national interest ought to be cultivated in Great Britain, is in fact being adequately cultivated in the universities. . . .
>
> In the view of the Vice-Chancellors, therefore, the universities may properly be expected not only individually to make proper use of the resources entrusted to them, but collectively to devise and execute policies calculated to serve the national interest. And in that task, both individually and collectively, they will be glad to have a greater measure of guidance from the Government than, until quite recent days, they have been accustomed to receive. . . .
>
> Hence the Vice-Chancellors would be glad if the University Grants Committee were formally authorised and equipped to undertake surveys of all main fields of university activity designed to secure that as a whole universities are meeting the whole range of national need for higher teaching and research. . . .

We meet here again with a passionate desire for accepting collective organisation for cultural activities, though these actually depend for their vigorous development on the initiative of individuals adjusting themselves to the advances of their rivals and guided by a cultural opinion in seeking support, be it public or private. It is true that competition between universities was getting increasingly concentrated on gaining the approval of the Treasury, and that its outcome came to determine to a considerable extent the framework within which the several universities could operate. But the most important administrative decisions, which determine the work of universities, as for example the selection of candidates for new vacancies, remained free and not arranged collectively by universities, but by competition between them. For they cannot be made otherwise. The Vice-Chancellors' memorandum has, in consequence, made no impression on the life of the universities and is, by this time, pretty well forgotten by the few who had ever seen it.[2]

WE may sum up by saying that the movements for guiding science towards a more direct service of the public interest, as well as for coordinating the pursuit of science more effectively from a centre, have all petered out. Science continues to be conducted in British universities as was done before the movement for the social guidance of science ever started. And I believe that all scientific progress achieved in the Soviet Union was also due—as everywhere else—to the initiative of original minds, choosing their own problems and carrying out their investigation, according to their own lights.

---

[2] I have never heard the memorandum mentioned in the University of Manchester. I knew about it only from Sir Ernest Simon's article entitled 'An Historical University Document,' in *Universities Quarterly*, February 1947, p. 189. My quotations referring to the memorandum are taken from this article.

This does not mean that society is asked to subsidise the private intellectual pleasures of scientists. It is true that the beauty of a particular discovery can be fully enjoyed only by the expert. But wide responses can be evoked by the purely scientific interest of discovery. Popular response, overflowing into the daily press, was aroused in recent years in England and elsewhere by the astronomical observations and theories of Hoyle and Lovell, and more recently by Ryle, and the popular interest was not essentially different from that which these advances had for scientists themselves.

And this is hardly surprising, since for the last three hundred years the progress of science has increasingly controlled the outlook of man on the universe, and has profoundly modified (for better and for worse) the accepted meaning of human existence. Its theoretic and philosophic influence was pervasive.

Those who think that the public is interested in science only as a source of wealth and power are gravely misjudging the situation. There is no reason to suppose that an electorate would be less inclined to support science for the purpose of exploring the nature of things, than were the private benefactors who previously supported the universities. Universities should have the courage to appeal to the electorate, and to the public in general, on their own genuine grounds. Honesty should demand this at least. For the only justification for the pursuit of scientific research in universities lies in the fact that the universities provide an intimate communion for the formation of scientific opinion, free from corrupting intrusions and distractions. For though scientific discoveries eventually diffuse into all people's thinking, the general public cannot participate in the intellectual milieu in which discoveries are made. Discovery comes only to a mind immersed in its pursuit. For such work the scientist needs a secluded place among likeminded colleagues who keenly share his aims and sharply control his performances. The soil of academic science must be exterritorial in order to secure its control by scientific opinion.

THE existence of this paramount authority, fostering, controlling and protecting the pursuit of a free scientific inquiry, contradicts the generally accepted opinion that modern science is founded on a total rejection of authority. This view is rooted in a sequence of important historical antecedents which we must acknowledge here. It is a fact that the Copernicans had to struggle with the authority of Aristotle upheld by the Roman Church, and by the Lutherans invoking the Bible; that Vesalius founded the modern study of human anatomy by breaking the authority of Galen. Throughout the formative centuries of modern science, the rejection of authority was its battle-cry; it was sounded by Bacon, by Descartes and collectively by the founders of the Royal Society of London. These great men were clearly saying something that was profoundly true and important but we should take into account today, the sense in which they have meant their rejection of authority. They aimed at adversaries who have since been defeated. And although other adversaries may have arisen in their places, it is misleading to assert that science is still based on the rejection of any kind of authority. The more widely the republic of science extends over the globe, the more numerous become its members in each country and the greater the material resources at its command, the more clearly emerges the need for a strong and effective scientific authority to reign over this republic. When we reject today the interference of political or religious authorities with the pursuit of science, we must do this in the name of the established scientific authority which safeguards the pursuit of science.

Let it also be quite clear that what we have described as the functions of scientific authority go far beyond a mere confirmation of facts asserted by science. For one thing, there are no mere facts in science. A scientific fact is one that has been accepted as such by scientific opinion, both on the grounds of the evidence in favour of it, and because it appears sufficiently plausible in view of the current scientific conception of the nature of things. Besides, science is not a mere collection of facts, but a system of facts based on their scientific interpretation. It is this system that is endorsed by a scientific authority. And within this system this authority endorses a particular distribution of scientific interest intrinsic to the system; a distribution of interest established by the delicate value-judgments exercised by scientific opinion in sifting and rewarding current contributions to science. Science *is what it is*, in virtue of the way in which scientific authority constantly eliminates, or else recognises at various levels of merit, contributions offered to science. In accepting the authority of science, we accept the totality of all these value-judgments.

Consider, also, the fact that these scientific evaluations are exercised by a multitude of scientists, each of whom is competent to assess only a tiny fragment of current scientific work, so that no single person is responsible at first hand for the announcements made by science at any time. And remember that each scientist originally established himself as such by joining at some point a network of mutual appreciation extending far beyond his own horizon. Each such acceptance appears then as a submission to a vast range of value-judgments exercised over all the domains of science, which the newly accepted citizen of science henceforth endorses, although he knows hardly anything about their subject-matter. Thus, the standards of scientific merit are seen to be transmitted from generation to generation by the affiliation of individuals at a great variety of widely disparate points,

15

in the same way as artistic, moral or legal traditions are transmitted. We may conclude, therefore, that the appreciation of scientific merit too is based on a tradition which succeeding generations accept and develop as their own scientific opinion. This conclusion gains important support from the fact that the methods of scientific inquiry cannot be explicitly formulated and hence can be transmitted only in the same way as an art, by the affiliation of apprentices to a master. The authority of science is essentially traditional.

BUT this tradition upholds an authority which cultivates originality. Scientific opinion imposes an immense range of authoritative pronouncements on the student of science, but at the same time it grants the highest encouragement to dissent from them in some particular. While the whole machinery of scientific institutions is engaged in suppressing apparent evidence as unsound, on the ground that it contradicts the currently accepted view about the nature of things, the same scientific authorities pay their highest homage to discoveries which deeply modify the accepted view about the nature of things. It took eleven years for the quantum theory, discovered by Planck in 1900, to gain final acceptance. Yet by the time another thirty years had passed, Planck's position in science was approaching that hitherto accorded only to Newton. Scientific tradition enforces its teachings in general, for the very purpose of cultivating their subversion in the particular.

I have said this here at the cost of some repetition, for it opens a vista of analogies in other intellectual pursuits. The relation of originality to tradition in science has its counterpart in modern literary culture. ' Seldom does the word [tradition] appear except in a phrase of censure ', writes T. S. Eliot.[3] He then tells how our exclusive appreciation of originality conflicts with the true sources of literary merit actually recognised by us:

> We dwell with satisfaction upon the poet's difference from his predecessors, especially his immediate predecessors; we endeavour to find something that can be isolated in order to be enjoyed. Whereas if we approach a poet without this prejudice, we shall often find that not only the best, but the most individual parts of his work may be those in which the dead poets, his ancestors, assert their immortality most vigorously.[4]

Eliot has also said, in *Little Gidding*, that ancestral ideas reveal their full scope only much later, to their successors:

> And what the dead had no speech for, when living,
> They can tell you, being dead: the communication
> Of the dead is tongued with fire beyond the language of the living.

And this is as in science: Copernicus and Kepler told Newton where to find discoveries unthinkable to themselves.

[3] T. S. Eliot, *Selected Essays*, London (1941), p. 13.
[4] *Ibid*. p. 14.

AT this point we meet a major problem of political theory: the question whether a modern society can be bound by tradition. Faced with the outbreak of the French Revolution, Edmund Burke denounced its attempt to refashion at one stroke all the institutions of a great nation, and predicted that this total break with tradition must lead to a descent into despotism. In reply to this, Tom Paine passionately proclaimed the right of absolute self-determination for every generation. The controversy has continued ever since. It has been revived in America in recent years by a new defence of Burke against Tom Paine, whose teachings had hitherto been predominant. I do not wish to intervene in the American discussion, but I think I can sum up briefly the situation in England during the past 170 years. To the most influential political writers of England, from Bentham to John Stuart Mill, and recently to Isaiah Berlin, liberty consists in doing what one likes, provided one leaves other people free to do likewise. In this view there is nothing to restrict the English nation *as a whole* in doing with itself at any moment whatever it likes. On Burke's vision of ' a partnership of those who are living, those who are dead and those who are to be born ' these leading British theorists turn a blind eye. But practice is different. In actual practice it is Burke's vision that controls the British nation; the voice is Esau's but the hand is Jacob's.

The situation is strange. But there must be some deep reason for it, since it is much the same as that which we have described in the organisation of science. This analogy seems indeed to reveal the reason for this curious situation. Modern man claims that he will believe nothing unless it is unassailable by doubt; Descartes, Kant, John Stuart Mill and Bertrand Russell have unanimously taught him this. They leave us no grounds for accepting any tradition. But we see now that science itself can be pursued and transmitted to succeeding generations only within an elaborate system of traditional beliefs and values, just as traditional beliefs have proved indispensable throughout the life of society. What can one do then? The dilemma is disposed of by continuing to profess the right of absolute self-determination in *political theory* and relying on the guidance of tradition in *political practice*.

But this dubious solution is unstable. A modern dynamic society, born of the French Revolution, will not remain satisfied indefinitely with accepting, be it only *de facto*, a traditional framework as its guide and master. The French Revolution, which, for the first time in history, had set up a government resolved on the indefinite improvement of human society, is still present in us. Its most far-reaching aspirations were embodied in the ideas of socialism, which rebelled against the whole structure of society and demanded its total renewal. In the twentieth century this demand went into action in Russia in an upheaval exceeding by far the range of the French Revolution. The boundless claims of the Russian Revolution have evoked passionate responses throughout the world. Whether accepted as a fervent conviction or repudiated as a menace, the ideas of the Russian Revolution have challenged everywhere the traditional framework which modern society had kept observing in practice, even though claiming absolute self-determination in theory.

I HAVE described how this movement evoked among many British scientists a desire to give deliberate social purpose to the pursuit of science. It offended their social conscience that the advancement of science, which affects the interests of society as a whole, should be carried on by individual scientists pursuing their own personal interests. They argued that all public welfare must be safeguarded by public authorities and that scientific activities should therefore be directed by the government in the interest of the public. This reform should replace by deliberate action towards a declared aim the present growth of scientific knowledge intended as a whole by no one, and in fact not even known in its totality, except quite dimly, to any single person. To demand the right of scientists to choose their own problems, appeared to them petty and unsocial, as against the right of society deliberately to determine its own fate.

But have I not said that this movement has virtually petered out by this time? Have not even the socialist parties throughout Europe endorsed by now the usefulness of the market? Do we not hear the freedom and the independence of scientific inquiry openly demanded today even in important centres within the Soviet domain? Why renew this discussion when it seems about to lose its point?

My answer is that you cannot base social wisdom on political disillusion. The more sober mood of public life today can be consolidated only if it is used as an opportunity for establishing the principles of a free society on firmer grounds. What does our political and economic analysis of the Republic of Science tell us for this purpose?

It appears, at first sight, that I have assimilated the pursuit of science to the market. But the emphasis should be in the opposite direction. The self-coordination of independent scientists embodies a higher principle, a principle which is *reduced* to the mechanism of the market when applied to the production and distribution of material goods.

LET me sketch out briefly this higher principle in more general terms. The Republic of Science shows us an association of independent initiatives, combined towards an indeterminate achievement. It is disciplined and motivated

18

by serving a traditional authority, but this authority is dynamic; its continued existence depends on its constant self-renewal through the originality of its followers.

The Republic of Science is a Society of Explorers. Such a society strives towards an unknown future, which it believes to be accessible and worth achieving. In the case of scientists, the explorers strive towards a hidden reality, for the sake of intellectual satisfaction. And as they satisfy themselves, they enlighten all men and are thus helping society to fulfil its obligation towards intellectual self-improvement.

A free society may be seen to be bent in its entirety on exploring self-improvement—every kind of self-improvement. This suggests a generalisation of the principles governing the Republic of Science. It appears that a society bent on discovery must advance by supporting independent initiatives, coordinating themselves mutually to each other. Such adjustment may include rivalries and opposing responses which, in society as a whole, will be far more frequent than they are within science. Even so, all these independent initiatives must accept for their guidance a traditional authority, enforcing its own self-renewal by cultivating originality among its followers.

Since a dynamic orthodoxy claims to be a guide in search of truth, it implicitly grants the right to opposition in the name of truth—truth being taken to comprise here, for brevity, all manner of excellence that we recognise as the ideal of self-improvement. The freedom of the individual safeguarded by such a society is therefore—to use the term of Hegel—of a positive kind. It has no bearing on the right of men to do as they please; but assures them the right to speak the truth as they know it. Such a society does not offer particularly wide private freedoms. It is the cultivation of public liberties that distinguishes a free society, as defined here.

In this view of a free society, both its liberties and its servitudes are determined by its striving for self-improvement, which in its turn is determined by the intimations of truths yet to be revealed, calling on men to reveal them.

This view transcends the conflict between Edmund Burke and Tom Paine. It rejects Paine's demand for the absolute self-determination of each generation, but does so for the sake of its own ideal of unlimited human and social improvement. It accepts Burke's thesis that freedom must be rooted in tradition, but transposes it into a system cultivating radical progress. It rejects the dream of a society in which all will labour for a common purpose, determined by the will of the people. For in the pursuit of excellence it offers no part to the popular will and accepts instead a condition of society in which the public interest is known only fragmentarily and is left to be achieved as the outcome of individual initiatives aiming at fragmentary problems. Viewed through the eyes of socialism, this ideal of a free society is conservative and fragmented, and hence adrift, irresponsible, selfish, apparently chaotic. A free society conceived as a society of explorers is open to these charges, in the sense that they do refer to characteristic features of it. But if we recognise that these features are indispensable to the pursuit of social self-improvement we may be prepared to accept them as perhaps less attractive aspects of a noble enterprise.

These features are certainly characteristic of the proper cultivation of science and are present throughout society as it pursues other kinds of truth. They are, indeed, likely to become ever more marked, as the intellectual and moral endeavours to which society is dedicated, enlarge in range and branch out into ever new specialised directions. For this must lead to further fragmentation of initiatives and thus increase resistance to any deliberate total renewal of society.

CRITERIA FOR SCIENTIFIC CHOICE

ALVIN M. WEINBERG

I

As science grows, its demands on our society's resources grow. It seems inevitable that science's demands will eventually be limited by what society can allocate to it. We shall then have to make choices. These choices are of two kinds. We shall have to choose among different, often incommensurable, fields of science—between, for example, high-energy physics and oceanography or between molecular biology and science of metals. We shall also have to choose among the different institutions that receive support for science from the government—among universities, governmental laboratories and industry. The first choice I call scientific choice; the second, institutional choice. My purpose is to suggest criteria for making scientific choices—to formulate a scale of values which might help establish priorities among scientific fields whose only common characteristic is that they all derive support from the government.

Choices of this sort are made at every level both in science and in government. The individual scientist must decide what science to do, what not to do: the totality of such judgments makes up his scientific taste. The research director must choose which projects to push, which to kill. The government administrator must decide not only which efforts to support; he must also decide whether to do a piece of work in a university, a national laboratory, or an industrial laboratory. The sum of such separate decisions determines our policy as a whole. I shall be concerned mainly with the broadest scientific choices: how should government decide between very large fields of science, particularly between different branches of basic science? The equally important question of how government should allocate its support for basic research among industry, governmental laboratories, and universities will not be discussed here.

II

Most of us like to be loved; we hate to make choices, since a real choice alienates the party that loses. If one is rich—more accurately, if one is growing richer—choices can be avoided. Every administrator knows that his job is obviously unpleasant only when his budget has been cut. Thus the urgency for making scientific or institutional choices has in the main been

This article from *Minerva*, I, 2 (Winter, 1963), pp. 159–171.

ignored both in the United States and elsewhere because the science budget has been expanding so rapidly: the United States government spent \$1,600,000,000 on research and development in 1950, \$9,000,000,000 in 1960, \$14,000,000,000 (including space) in 1962.

Though almost all agree that choices will eventually have to be made, some well-informed observers insist that the time for making the choices is far in the future. Their arguments against making explicit choices have several main threads. Perhaps most central is the argument that since we do not make explicit choices about anything else, there is no reason why we should make them in science. Since we do not explicitly choose between support for farm prices and support for schools, or between highways and foreign aid, why should we single out science as the guinea pig for trying to make choices? The total public activity of our society has always resulted from countervailing pressures, exerted by various groups representing professional specialities, or local interests, or concern for the public interest. The combination that emerges as our Federal budget is not arrived at by the systematic application of a set of criteria: even the highest level of authority, in the United States, the President, who must weigh conflicting interests in the scale of the public interest, is limited in the degree to which he can impose an overall judgment by the sheer size of the budget if by nothing else. But because we have always arrived at an allocation by the free play of countervailing pressures this does not mean that such free interplay is the best or the only way to make choices. In any case, even if our choices remain largely implicit rather than explicit, they will be more reasonable if persons at every level, representing every pressure group, try to understand the larger issues and try to mitigate sectional self-interest with concern for broader issues. The idea of conflicting and biased claims being adjudicated at one fell swoop by an all-knowing supreme tribunal is a myth. It is much better that the choices be decentralised and that they reflect the concern for the larger interest. For this reason alone philosophic debate on the problems of scientific choice should lead to a more rational allocation of our resources.

A second thread in the argument of those who refuse to face the problem of scientific choice is that we waste so much on trivialities—on smoking, on advertising, on gambling—that it is silly to worry about expenditures of the same scale on what is obviously a more useful social objective—the increase of scientific knowledge. A variant of this argument is that with so much unused steel capacity or so many unemployed, we cannot rightly argue that we cannot afford a big cyclotron or a large manned-space venture.

Against these arguments we would present the following considerations on behalf of a rational scientific policy. At any given instant, only a certain fraction of our society's resources goes to science. To insist or imply that the *summum bonum* of our society is the pursuit of science and that therefore all other activities of the society are secondary to science—that unused capacity in the steel mills should go to ' Big Science ' rather than a large-scale housing programme—is a view that might appeal strongly to the scientific community. It is hardly likely to appeal so strongly to the much larger part of society that elects the members of the legislature, and to whom, in all probability, good houses are more important than good science. Thus, as a practical matter we cannot really evade the problem of scientific choice. If those actively engaged in science do not make choices, they will be made anyhow by the Congressional Appropriations Committees and by the Bureau of the Budget, or corresponding bodies in other governments. Moreover, and perhaps more immediately, even if we are not limited by money, we shall be limited by the availability of truly competent men. There is already evidence that our ratio of money to men in science is too high, and that in some parts of science we have gone further more quickly than the number of really competent men can justify.

III

Our scientific and governmental communities have evolved institutional and other devices for coping with broad issues of scientific choice. The most important institutional device in the United States is the President's Science Advisory Committee, with its panels and its staff in the Office of Science and Technology. This body and its panels help the Bureau of the Budget to decide what is to be supported and what is not to be supported. The panel system, however, suffers from a serious weakness. Panels usually consist of specialised experts who inevitably share the same enthusiasms and passions. To the expert in oceanography or in high energy physics, nothing seems quite as important as oceanography or high energy physics. The panel, when recommending a programme in a field in which all its members are interested, invariably argues for better treatment of the field—more money, more people, more training. The panel system is weak insofar as judge, jury, plaintiff and defendant are usually one and the same.

The panel is able to judge how competently a proposed piece of research is likely to be carried out: its members are all experts and are likely to know who are the good research workers in the field. But just because the panel is composed of experts, who hold parochial viewpoints, the panel is much less able to place the proposal in a broader perspective and to say whether the research proposal is of much interest to the rest of science. We can answer the question ' how ' within a given frame of reference; it is impossible to answer ' why ' within the same frame of reference. It would therefore seem that the panel system could be improved if representatives, not only of the field being judged but also representatives of neighbouring fields, sat on every panel judging the merits of a research proposal. A panel judging high energy physics should have some people from low energy physics; a panel judging low energy physics should have some people from nuclear energy; a panel judging nuclear energy should have some people from conventional energy; and so on. I should think that advice from panels so constituted would be tempered by concern for larger issues; in particular, the support of a proposed research project would be viewed from the larger perspective of the relevance of that research to the rest of science.

In addition to panels or the bodies like the President's Science Advisory Committee as organisational instruments for making choices, the scientific community has evolved an empirical method for establishing scientific priorities, that is, for deciding what is important in science and what is not important. This is the scientific literature. The process of self-criticism, which is integral to the literature of science, is one of the most characteristic features of science. Nonsense is weeded out and held up to ridicule in the literature, whereas what is worthwhile receives much sympathetic attention. This process of self-criticism embodied in the literature, though implicit, is nonetheless real and highly significant. The existence of a healthy, viable scientific literature in itself helps assure society that the science it supports is valid and deserving of support. This is a most important, though little recognised, social function of the scientific literature.

As an arbiter of scientific taste and validity, scientific literature is beset with two difficulties. First, because of the information explosion, the literature is not read nearly as carefully as it used to be. Nonsense is not so generally recognised as such, and the standards of self-criticism, which are so necessary if the scientific literature is to serve as the arbiter of scientific taste, are inevitably looser than they once were.

Second, the scientific literature in a given field tends to form a closed universe; workers in a field, when they criticise each other, tend to adopt the same unstated assumptions. A referee of a scientific paper asks whether the paper conforms to the rules of the scientific community to which both referee and author belong, not whether the rules themselves are valid. So to speak, the editors and authors of a journal in a narrowly specialised field are all tainted with the same poison. As Einstein said, ' Eigener Dreck stinkt nicht.' [1]

Can a true art of scientific criticism be developed, *i.e.*, can one properly criticise a field of science beyond the kind of criticism that is inherent in the literature of the field? Mortimer Taube in *Computers and Common*

---

[1] As quoted by Dyson, Freeman J., in a review of Sweber, S.S., *Mesons and Fields*, in *Physics Today*, IX (May, 1956), pp. 32–34.

Sense[2] insists that such scientific criticism is a useful undertaking, and that, by viewing a field from a somewhat detached point of view, it is possible to criticise a field meaningfully, even to the point of calling the whole activity fraudulent, as he does in the case of non-numerical uses of computers. I happen to believe that Taube does not make a convincing case in respect to certain non-numerical uses of computers, such as language translation. Yet I have sympathy for Dr. Taube's aims—that, with science taking so much of the public's money, we must countenance, even encourage, discussion of the relative validity and worthwhileness of the science which society supports.

IV

I believe that criteria for scientific choice can be identified. In fact, several such criteria already exist; the main task is to make them more explicit. The criteria can be divided into two kinds: internal criteria and external criteria. Internal criteria are generated within the scientific field itself and answer the question: How well is the science done? External criteria are generated outside the scientific field and answer the question: Why pursue this particular science? Though both are important, I think the external criteria are the more important.

Two internal criteria can be easily identified: (1) Is the field ready for exploitation? (2) Are the scientists in the field really competent? Both these questions are answerable only by experts who know the field in question intimately, and who know the people personally. These criteria are therefore the ones most often applied when a panel decides on a research grant: in fact, the primary question in deciding whether to provide governmental support for a scientist is usually: How good is he?

I believe, however, that it is not tenable to base our judgments entirely on internal criteria. As I have said, we scientists like to believe that the pursuit of science as such is society's highest good, but this view cannot be taken for granted. For example, we now suffer a serious shortage of medical practitioners, probably to some extent because many bright young men who would formerly have gone into medical practice now go into biological research; government support is generally available for post-graduate study leading to the Ph.D. but not for study leading to the medical degree. It is by no means self-evident that society gains from more biological research and less medical practice. Society does not *a priori* owe the scientist, even the good scientist, support any more than it owes the artist or the writer or the musician support. Science must seek its support from society on grounds other than that the science is carried out competently and that it is ready for

[2] (New York: Columbia University Press, 1961).

exploitation; scientists cannot expect society to support science because scientists find it an enchanting diversion. Thus, in seeking justification for the support of science, we are led inevitably to consider external criteria for the validity of science—criteria external to science, or to a given field of science.

V

Three external criteria can be recognised: technological merit, scientific merit and social merit. The first is fairly obvious: once we have decided, one way or another, that a certain technological end is worthwhile, we must support the scientific research necessary to achieve that end. Thus, if we have set out to learn how to make breeder reactors, we must first measure painstakingly the neutron yields of the fissile isotopes as a function of energy of the bombarding neutron. As in all such questions of choice, it is not always so easy to decide the technological relevance of a piece of basic research. The technological usefulness of the laser came after, not before, the principle of optical amplification was discovered. But it is my belief that such technological bolts from the scientific blue are the exception, not the rule. Most programmatic basic research can be related fairly directly to a technological end at least crudely if not in detail. The broader question as to whether the technological aim itself is worthwhile must be answered again partly from within technology through answering such questions as: Is the technology ripe for exploitation? Are the people any good? Partly from outside technology by answering the question: Are the social goals attained, if the technology succeeds, themselves worthwhile? Many times these questions are difficult to answer, and sometimes they are answered incorrectly: for example, the United States launched an effort to control thermonuclear energy in 1952 on a rather large scale because it was thought at the time that controlled fusion was much closer at hand than it turned out to be. Nevertheless, despite the fact that we make mistakes, technological aims are customarily scrutinised much more closely than are scientific aims; at least we have more practice discussing technological merit than we do scientific merit.

VI

The criteria of scientific merit and social merit are much more difficult: scientific merit because we have given little thought to defining scientific merit in the broadest sense, social merit because it is difficult to define the values of our society. As I have already suggested, the answer to the question: Does this broad field of research have scientific merit? cannot be answered within the field. The idea that the scientific merit of a field can be judged better from the vantage point of the scientific fields in which it is embedded than from the point of view of the field itself is implicit in the following quotation from the late John von Neumann: ' As a mathematical discipline travels far from its empirical source, or still more, if it is a second and third generation only indirectly inspired by ideas coming from reality, it is beset with very grave dangers. It becomes more and more pure aestheticising, more and more purely *l'art pour l'art*. This need not be bad if the field is surrounded by correlated subjects which still have closer empirical connections or if the discipline is under the influence of men with an exceptionally well-developed taste. But there is a grave danger that the subject will develop along the line of least resistance, that the stream, so far from its source, will separate into a multitude of insignificant branches, and that the discipline will become a disorganised mass of details and complexities. In other words, at a great distance from its empirical source, or after much " abstract " inbreeding, a mathematical subject is in danger of degeneration. At the inception the style is usually classical; when it shows signs of becoming baroque, then the danger signal is up.' [3]

I believe there are any number of examples to show that von Neumann's observation about mathematics can be extended to the empirical sciences. *Empirical* basic sciences which move too far from the neighbouring sciences in which they are embedded tend to become ' baroque '. Relevance to neighbouring fields of science is, therefore, a valid measure of the scientific merit of a field of basic science. In so far as our aim is to increase our grasp and understanding of the universe, we must recognise that some areas of basic science do more to round out the whole picture than do others. A field in which lack of knowledge is a bottleneck to the understanding of other fields deserves more support than a field which is isolated from other fields. This is only another way of saying that, ideally, science is a unified structure and that scientists, in adding to the structure, ought always to strengthen its unity. Thus, the original motivation for much of high-energy physics is to be sought in its elucidation of low-energy physics, or the strongest and most exciting motivation for measuring the neutron capture cross sections of the elements lies in the elucidation of the cosmic origin of the elements. Moreover, the discoveries which are acknowledged to be the most important scientifically, have the quality of bearing strongly on the scientific disciplines around them. For example, the discovery of X-rays was important partly because it extended the electromagnetic spectrum but, much more, because it enabled us to see so much that we had been unable to see. The word ' fundamental ' in basic science, which is often used as a synonym for ' important ', can be partly paraphrased into ' relevance to neighbouring areas

[3] Heywood, R. B. (ed.), *The Works of the Mind* (University of Chicago Press, 1947), p. 196.

of science '. I would therefore sharpen the criterion of scientific merit by proposing that, other things being equal, *that field has the most scientific merit which contributes most heavily to and illuminates most brightly its neighbouring scientific disciplines.* This is the justification for my previous suggestion about making it socially acceptable for people in *related* fields to offer opinions on the scientific merit of work in a given field. In a sense, what I am trying to do is to extend to basic research a practice that is customary in applied science: a project director trying to get a reactor built on time is expected to judge the usefulness of component development and fundamental research which bears on his problems. He is not always right; but his opinions are usually useful both to the researcher and to the management disbursing the money.

VII

I turn now to the most controversial criterion of all—social merit or relevance to human welfare and the values of man. Two difficulties face us when we try to clarify the criterion of social merit: first, who is to define the values of man, or even the values of our own society; and second, just as we shall have difficulty deciding whether a proposed research helps other branches of science or technology, so we will have even greater trouble deciding whether a given scientific or technical enterprise indeed furthers our pursuit of social values, even when those values have been identified. With some values we have little trouble: adequate defence, or more food, or less sickness, for example, are rather uncontroversial. Moreover, since such values themselves are relatively easy to describe, we can often guess whether a scientific activity is likely to be relevant, if not actually helpful, in achieving the goal. On the other hand, some social values are much harder to define: perhaps the most difficult is national prestige. How do we measure national prestige? What is meant when we say that a man on the moon enhances our national prestige? Does it enhance our prestige more than, say, discovering a polio vaccine or winning more Nobel Prizes than any other country? Whether or not a given achievement confers prestige probably depends as much on the publicity that accompanies the achievement as it does on its intrinsic value.

Among the most attractive social values that science can help to achieve is international understanding and cooperation. It is a commonplace that the standards and loyalties of science are trans-national. A new element has recently been injected by the advent of scientific research of such costliness that now it is prudent as well as efficient to participate in some form of international cooperation. The very big accelerators are so expensive that international laboratories such as CERN at Geneva are set up to enable several countries to share costs that are too heavy for them to bear separately. Even if we were not committed to improving international relations we would be impelled to cooperate merely to save money.

Bigness is an advantage rather than a disadvantage if science is to be used as an instrument of international cooperation: a $500,000,000 cooperative scientific venture—such as the proposed 1,000 Bev intercontinental accelerator—is likely to have more impact than a $500,000 Van de Graaff machine. The most expensive of all scientific or quasi-scientific enterprises —the exploration of space—is, from this viewpoint, the best-suited instrument for international cooperation. The exchange between President Kennedy and Chairman Khrushchev concerning possible increased cooperation in space exploration seems to have been well received and, one hopes, will bear ultimate fruit.

VIII

Having set forth these criteria and recognising that judgments are fraught with difficulty, I propose to assess five different scientific and technical fields, in the light of these. The five fields I choose are molecular biology, high-energy physics, nuclear energy, manned-space exploration, and the behavioural sciences. Two of these fields, molecular biology and high-energy physics, are, by any definition, basic sciences; nuclear energy is applied science, the behavioural sciences are a mixture of both applied and basic science. Manned exploration of space, though it requires the tools of science and is regarded in the popular mind as being part of science, has not yet been proved to be more than quasi-scientific, at best. The fields which I choose are incommensurable: how can one measure the merit of behavioural sciences and nuclear energy on the same scale of values? Yet the choices between scientific fields will eventually have to be made whether we like it or not. Criteria for scientific choice will be most useful only if they *can* be applied to seemingly incommensurable situations. The validity of my proposed criteria depends on how well they can serve in comparing fields that are hard to compare.

Of the scientific fields now receiving public support, perhaps the most successful is molecular biology. Hardly a month goes by without a stunning success in molecular biology being reported in the *Proceedings of the National Academy of Sciences*. The most recent has been the cracking by Nierenberg and Ochoa of the code according to which triples of bases determine specific amino acids in the living proteins. Here is a field which rates the highest grades as to its ripeness for exploitation and competence of its workers. It is profoundly important for large stretches of other biological sciences—genetics, cytology, microbiology—and, therefore, according to my criterion, must be graded A + for its scientific merit. It also must be given a very high grade in social merit, and probably in technological (that is, medical) merit—more than, say, taxonomy or topology. Molecular biology is the most fundamental of all the biological sciences. With understanding of the manner of transmission of genetic information ought to come the insights necessary for the solution of such problems as cancer, birth defects, and viral diseases. Altogether, molecular biology ought, in my opinion, to receive as much public support as can possibly be pumped into it; since money is not limiting its growth, many more post-graduate students and research fellows in molecular biology ought to be subsidised so that the attack on this frontier can be expanded as rapidly as possible.

The second field is high-energy physics. This field of endeavour originally sought as its major task to understand the nuclear force. In this it has been only modestly successful; instead, it has opened an undreamed-of subnuclear world of strange particles and hyperons, a world in which mirror images are often reversed. The field has no end of interesting things to do, it knows how to do them, and its people are the best. Yet I would be bold enough to argue that, at least by the criteria which I have set forth— relevance to the sciences in which it is embedded, relevance to human affairs, and relevance to technology—high-energy physics rates poorly. The nuclear forces are not being worked on very directly—the world of subnuclear particles seems to be remote from the rest of the physical sciences. Aside from the brilliant resolution of the τ-particle paradox, which led to the overthrow of the conservation of parity, and the studies of mesic atoms (the latter of which is not done at *ultra*-high energy), I know of few discoveries in ultra-high-energy physics which bear strongly on the rest of science. (This view would have to be altered if machines such as the Argonne Zero Gradient Synchrotron were exploited as very strong, pulsed sources of neutrons for study of neutron cross sections.) As for its bearing on human welfare and on technology, I believe it is essentially nil. These two low grades would not bother me if high-energy physics were cheap. But it is terribly expensive— not so much in money as in highly qualified people, especially those brilliant talents who could contribute so ably to other fields which contribute much more to the rest of science and to humanity than does high-energy physics. On the other hand, if high-energy physics could be made a vehicle for international cooperation—if the much-discussed intercontinental 1,000 Bev accelerator could indeed be built as a joint enterprise between East and West —the expense of high-energy physics would become a virtue, and the enterprise would receive a higher grade in social merit than I would now be willing to assign to it.

Third is nuclear energy. This being largely an applied effort, it is very relevant to human welfare. We now realise that in the residual uranium and thorium of the earth's crust, mankind has an unlimited store of energy— enough to last for millions of years; and that with an effort of only one-tenth of our manned-space effort we could, within ten or fifteen years, develop the reactors which would tap this resource. Only rarely do we see ways of *permanently* satisfying one of man's major needs—in this case energy. In high-conversion ratio nuclear reactors we have such means, and we are close to their achievement. Moreover, we begin to see ways of applying very large reactors of this type to realise another great end, the economic desalination of the ocean. Thus, the time is very ripe for exploitation. Nuclear energy rates so highly in the categories of technical and social merit and timeliness that I believe it deserves strong support, even if it gets very low marks in the other two categories—its personnel and its relationship to the rest of science. Suffice it to say that in my opinion the scientific workers in the field of nuclear energy are good and that nuclear energy in its basic aspects has vast ramifications in other scientific fields.

Next on the list are the behavioural sciences—psychology, sociology, anthropology, economics. The workers are of high quality; the sciences are significantly related to each other, they are deeply germane to every aspect of human existence. In these respects the sciences deserve strong public support. On the other hand, it is not clear to me that the behavioural scientists, on the whole, see clearly how to attack the important problems of their sciences. Fortunately, the total sum involved in behavioural science research is now relatively tiny—as it well must be when what are lacking are deeply fruitful, and generally accepted, points of departure.

Finally, I come to manned-space exploration. The personnel in the programme are competent and dedicated. With respect to ripeness for exploitation, the situation seems to me somewhat unclear. Our ' hardware ' is in good shape, and we can expect it to get better—bigger and more reliable boosters, better communication systems, etc. What is not clear is the human being's tolerance of the space environment. I do not believe that either the hazards of radiation or of weightlessness are sufficiently explored yet positively to guarantee success in our future manned-space ventures.

The main objection to spending so much manpower, not to say money, on manned-space exploration is its remoteness from human affairs, not to say the rest of science. In this respect space (the exploration of very large distances) and high-energy physics (the exploration of very small distances) are similar, though high-energy physics has the advantage of greater scientific validity. There are some who argue that the great adventure of man into space is not to be judged as science, but rather as a quasi-scientific enterprise, justified on the same grounds as those on which we justify other non-scientific national efforts. The weakness of this argument is that space requires many, many scientists and engineers, and these are badly needed for such matters as clarifying our civilian defence posture or, for that matter, working out the technical details of arms control and foreign aid. If space is ruled to be non-scientific, then it must be balanced against other non-scientific expenditures like highways, schools or civil defence. If we do space-research because of prestige, then we should ask whether we get more prestige from a man on the moon than from successful control of the water-logging problem in Pakistan's Indus Valley Basin. If we do space-research because of its military implications, we ought to say so—and perhaps the military justification, at least for developing big boosters, is plausible, as the Soviet experience with rockets makes clear.

IX

The main weight of my argument is that the most valid criteria for assessing scientific fields come from without rather than from within the scientific discipline that is being rated. This does not mean that only those scientific fields deserve priority that have high technical merit or high social merit. Scientific merit is as important as the other two criteria, but, as I have argued, scientific merit must be judged from the vantage point of the scientific fields in which each field is embedded rather than from that of the field itself. If we support science in order to maximise our knowledge of the world around us, then we must give the highest priority to those scientific endeavours that have the most bearing on the rest of science.

The rather extreme view which I have taken presents difficulties in practice. The main trouble is that the bearing that one science has on another science so often is not appreciated until long after the original discoveries have been made. Who was wise enough, at the time Purcell and Bloch first discovered nuclear magnetic resonance, to guess that the method would become an important analytical tool in biochemistry? Or how could one have guessed that Hahn and Strassmann's radiochemical studies would have led to nuclear energy? And indeed, my colleagues in high-energy physics predict that what we learn about the world of strange particles will in an as yet undiscernible way teach us much about the rest of physics, not merely much about strange particles. They beg only for time to prove their point.

To this argument I say first that choices are always hard. It would be far simpler if the problem of scientific choice could be ignored, and possibly in some future millennium it can be. But there is also a more constructive response. The necessity for scientific choice arises in ' Big Science ', not in

' Little Science '. Just as our society supports artists and musicians on a small scale, so I have no objection to—in fact, I strongly favour—our society supporting science that rates zero on all the external criteria, provided it rates well on the internal criteria (ripeness and competence) and provided it is carried on on a relatively small scale. It is only when science really does make serious demands on the resources of our society—when it becomes ' Big Science '—that the question of choice really arises.

At the present time, with our society faced with so much unfinished and very pressing business, science can hardly be considered its major business. For scientists as a class to imply that science can, at this stage in human development, be made the main business of humanity is irresponsible— and, from the scientist's point of view, highly dangerous. It is quite conceivable that our society will tire of devoting so much of its wealth to science, especially if the implied promises held out when big projects are launched do not materialise in anything very useful. I shudder to think what would happen to science in general if our manned-space venture turned out to be a major failure, if it turned out, for example, that man could not withstand the re-entry deceleration forces after a long sojourn in space. It is as much out of a prudent concern for their own survival, as for any loftier motive, that scientists must acquire the habit of scrutinising what they do from a broader point of view than has been their custom. To do less could cause a popular reaction which would greatly damage mankind's most remarkable intellectual attainment—Modern Science—and the scientists who created it and must carry it forward.

THE DISTRIBUTION OF SCIENTIFIC EFFORT

C. F. CARTER

I

THE work of scientists and technologists can be classified in several ways. For instance, it can be described by the fundamental discipline involved— as the work of chemists, physicists, engineers, biologists, or in more detail as the work of astrophysicists, chemical engineers, and so on. It can be described as 'pure research', 'applied research', and 'other scientific and technological work' (*e.g.*, process control). It can be classified as research, design, development, production, administration. Any such method of division carries with it the implied question—is the division (in the circumstances of a particular country and a particular time) the 'right' division? This leads in turn to further questions. By what criteria can we judge whether the distribution of scientific effort is 'right'? What, indeed, do we mean by the 'rightness' of a distribution of effort?

These are questions with important implications for public policy. It is certainly possible to some degree to alter the distribution of types of scientist and technologist by manipulating the educational system; the creation and expansion of colleges of advanced technology in Britain are efforts to do just that. It is possible to a considerable degree to alter the points at which scientific effort is applied, by giving or withholding research contracts or grants or aids to development. But I do not think that the determination of public policy has often had the benefit of clear thinking about the distribution of scientific effort. The purpose of this article is not to provide this clear thinking, but in a more humble way to try to define the questions which ought to be asked. I shall draw my examples mainly from British experience.

II

First, it is necessary to deal with a question which, though it is often asked, is almost certainly not the right way to start this inquiry. This is the question: Where are the gaps in our national effort? It conceives of the government as a prudent gardener, who observes that some of his flower-beds are flourishing and others bare for lack of water, and who goes out with the watering-can to stimulate growth in the bare places. But water for gardens is not usually (in Britain) in short supply. There is no real choice

This article from *Minerva*, I, 2 (Winter, 1963), pp. 172–181.

in the use of scarce assets to be made; all that is needed is a little observation, thought and trouble from the gardener.

Scientists, on the other hand, are in short supply; the enumeration of gaps may stimulate thought about the balance of effort, but it cannot rightly carry the implication that gaps should not exist, for it is doubtful if even the largest and richest countries can now command the scientific resources needed to cultivate all the numerous and multiplying specialisms of science and technology. The Advisory Council on Scientific Policy, in its report for 1959–60 of a study on 'the balance of scientific effort',[1] rightly says that ' our resources hardly permit us to be active at all points at once ', but it goes on to talk about ' subjects which have been relatively neglected ' and about ' too many fallow fields '—leaving a mild suspicion that the eminent members of the Council regarded the existence of a gap as *prima facie* evidence that it ought to be filled. If we lived in a really small country (say, Sweden or Ireland) we could, of course, have no such illusion.

The gap-filling mentality is on occasion accompanied by a far less excusable illusion (though one which is readily to be found in the United States and Russia), namely, that it is discreditable if one's home country is not simultaneously a world leader in all lines of science and technology. The discovery that another country has a lead in a particular area of science is, of course, no sufficient reason for increasing the effort devoted to that area, as though one were in danger of relegation to a lower division in some intellectual football league. Indeed, it could be argued that evidence of successful scientific effort in another country is a good reason for sending a telegram of congratulation, and turning one's attention to some different area of the subject. In the international community of scientists there might be an advantage in specialisation.

This, however, is not an argument to be pressed too far. The channels by which scientific knowledge flows run for the most part between specialists in the same subject, and thence outwards to fertilise other subjects or to find development and application. If the world concentrated its effort for a particular science on a particular centre, it might indeed achieve massive results; but unfortunately no other centre or country would contain people capable of understanding and assimilating these results. We have to set against the argument for specialisation, therefore, an argument that there should be a reasonable spread of scientific interest in each separate area. For this purpose a ' separate area ' must often be identified with a nation-state, with its separate language, institutions and commercial links, making it much easier for ideas to flow within its area than across its boundaries. Thus the argument for a spread of scientific interests, applied to small or

[1] Cmnd. 1167 (London: H.M. Stationery Office, 1960).

poor countries, may imply that they need an apparently wasteful diversity of scientific interest, in order that they may have 'receiving stations' for the flow of ideas from outside.

Here, then, is a first principle: that though a country should not strive to fill every gap, it needs a large enough spread of scientific interest to give it a reasonable understanding of the broad lines of scientific advance elsewhere in the world. But though this tells us a little about 'where', it tells us nothing about 'how much'. The aggregate scientific effort of a country is a function of the number of scientists and technologists (of various kinds) which it contains, and of the effectiveness of their use. I have suggested elsewhere [2] that we understand very little about what constitutes 'effective use' of a country's stock of trained ability, and that there is need for research on many aspects of this problem. For instance, would we increase the productiveness of scientists by a greater supply of ancillary staff of lower intelligence or training? If so, how far should the use of ancillary staff be pushed? But these are issues which, in this article, I propose to dodge; I will assume that it is national policy to have full employment of its scientists, and to use their skill as fully as our present understanding of the 'use of skill' allows.

To begin with I will take no account of the use of scientists in production or administration, and simply consider the division of scientific effort between pure and applied research (the latter here taken to include development). The difference between a 'pure' and an 'applied' subject is one of motive rather than of the nature of the work; applied research finds its justification in the consequences of its application, pure research in the inherent interest of solving the mysteries of nature. But, of course, pure research often finds later application (however useless it may have seemed to begin with), and therefore there is a tendency to judge it by a standard other than its true motive. It is still often believed that British scientists are good at fundamental discovery, but poor at application (and, oddly enough, the same belief is held by several other countries about their own scientists). But if one uses this as an argument for reducing the effort in pure science, so as to increase that in applied science and development, one has, in effect, treated pure science as the first stage in a productive process which ought to be in balance, and thus one has judged it by its ultimate usefulness rather than by its inherent interest.

It can well be asked, however, whether there is any particular hurry about exploring the mysteries of nature. Does it matter whether a man lands on the moon in 1970 or in two centuries' time? Are not the arguments for haste mostly derived from a hope of advantage in the applied field of military

[2] 'The Economic Use of Brains', *Economic Journal*, LXXII (March, 1962), 285, pp. 1–11.

36

technology? Indeed, our precipitate enthusiasm for pure scientific discovery is running far beyond our political skill or social understanding. Perhaps it may be illuminating to consider the total effort of pure research in two parts. One part would be capable of justification by its ultimate usefulness in application, *as well as* by its inherent interest; the other part would be a work of supererogation, justifiable only by the pure joy of discovery. A nation which chose to undertake pure research beyond what can be given a broad economic justification by its ultimate application would have regarded intellectual discovery as more important than an increment of wealth. ' How much better is it to get wisdom than gold! And to get understanding rather to be chosen than silver! ' [3]

In point of fact, however, it would be difficult to prove from the British record that the total scale of pure research exceeds what can be given an economic justification. It is not enough to produce cases of British discoveries in pure science which have received their application abroad; naturally, if the streams of knowledge flow at all between nations, such things will happen. Nor is it enough to show that there are well-developed areas of knowledge which look ripe for application, but have not received it; for it is equally possible to find applications which are held up for lack of fundamental knowledge in related fields. The impression that we are (from an economic point of view) disproportionately strong in pure science is probably derived from the fact that pure science has the greater social prestige.

III

Thus far I have been dealing with an extremely general (and imprecise) distinction between the pure and the applied. But even if this balance cannot be proved inappropriate, it may still be true that there are too few engineers relative to chemists, or too few design engineers relative to the total of engineers. Is there any hope of finding principles by which to judge the detail of the distribution of scientific effort? Or must it be left simply as the consequence of accidents of decision? It depends at present on the decisions of students about specialisation at school and university, and the decisions made in the past about the educational facilities to be provided; both sets of decisions being made without any clear idea of the national need.

It seems to me that, though this is not entirely an economic matter, it is only from economics that any guidance will at present be obtained. In other words, it is possible to give some sort of answer to the question: What kind of distribution of scientific effort will most effectively increase the flow of wealth? But I see no means of finding out what distribution of effort will

[3] *Proverbs*, xvi, v. 16.

maximise human happiness, or maximise the joys of discovery. How can one compare the interest or excitement of a discovery about cell structure in living organisms with that of a discovery about magnetic fields in the Milky Way? How can one estimate whether, when the balance is finally struck, the fears of destruction which are founded on discoveries in nuclear physics will exceed the satisfactions of a world freed from dangers of energy shortage? But though, in consequence, I turn to propose an economic analysis, this is not because I believe that the economic criterion should always prevail. The point is rather that, if we do not have an economic policy about the distribution of scientific resources, we are unlikely to have any policy at all. If we begin by trying to find out what distribution will most effectively increase the flow of wealth, we can use what we find as a framework for further thinking. We may then decide that, because we attach a high utility to the satisfaction of our curiosity, or because we wish to glorify God in the study of an aspect of His creation, we will, *at the expense of the flow of material wealth,* devote more resources (say) to space research. Such a decision could reasonably be made, because it is a clear choice between alternatives; but without an economic policy we have no point of departure, except perhaps some amalgam of the preferences of scientists (who are likely to be in favour of the further development of their own specialities).

Let us look still at the use of scientists in research and development. Their function in the productive system is to provide an immaterial input of ideas. It is often our habit to concentrate attention on the material inputs to production, and on the labour of human hands; but, as our wealth advances, the immaterial inputs of scientific knowledge, artistic design and organisational skill become increasingly important. The relation between the input of ideas and the flow of wealth is, however, of a special kind. If we talk of an input of sausage-meat helping to produce an output of sausages, we are relating one flow to another. The scientific knowledge embodied in a flow of sausages, however, is a stock; it is the current state of knowledge about preservatives, cases, filling techniques and so forth. It is not a capital stock of the same kind as a machine, for a machine is used up or depreciated in the course of production, and its value lies in its power to yield services over a limited period of years. The stock of knowledge does not wear out in being used, though it may be superseded by later knowledge.

It follows that we shall be asking the wrong question if we seek a relationship between the *flow* of research and development of different kinds and the current output of the country. The relationship is between the stock of knowledge and the flow of wealth; and the stock of knowledge is created by the *cumulative* expenditures on research and development in past years, and on the cumulative inflow of ideas from elsewhere. In saying this I do not ignore the fact that accretions of knowledge come in random jumps of different sizes, and that it does not follow that, in a particular area of science, doubling the expenditure on research will double (or increase at all) the probability of significant discovery. But this observation of randomness seems to me to be given too much weight as a fact of history, and it is not relevant to the determination of scientific policy, unless indeed it is used to show that it is impossible to have any policy at all. I reject this gloomy view, for it is surely rational to assume a broad relationship between the probability of discovery and the resources used in trying to achieve it—a relationship subject to random disturbance, but nevertheless likely to assert its dominance in the long run.

We must also take account of the fact that ideas arrive, not as a continuous flow, but in discrete quanta which are the result of particular projects of research and development. These projects are the survivors of a larger number which were started, but some of which failed to yield a useful result. Each project stretches over time, and often over a very considerable time. It follows that the quanta of knowledge which will be contributed to the stock in 1963 will for the most part result from projects started long before, and no alteration of the distribution of scientific resources in 1963 will have much effect on the stock of knowledge *in that year*.

When New Year's Day dawned in 1963 the enterprises of the British economy had a stock of knowledge and ideas, adopted, or available to be adopted—a stock derived by slow accretion from the earliest days of man's existence. The object of an economic policy is, then, to alter the uses of scientific resources, within the limits set by the available stock of scientific skills, so that *gradually* (as future projects yield their results) the flow of new knowledge may alter the stock of knowledge in a desirable way. This implies that the strategy for research and development needs to be decided, not in relation to the circumstances of 1963, nor even within the framework of a five-year plan such as that which is presumably to come from the National Economic Development Council. It involves a look ahead for ten, fifteen or twenty years; for it is only in a time as long as this that it is possible both to alter the use of scientific resources, and to bring a significant number of new projects to fruition.

But what is the 'desirable way' in which to alter the stock of knowledge? One must be careful of the obvious answers, such as 'to increase the productivity of labour'; in a country like China, with almost limitless reserves of labour, such an answer would be wrong. Emphasis on improving the productivity of labour, *e.g.*, by automation, would be appropriate in a country with a serious shortage of labour, but with ample land, capital and organising ability. We must look, in fact, to see what it is that is *over the long run* likely to be most important in holding back economic growth. It is possible to argue that Britain is short of labour of certain skills, short of capital equipment (*e.g.*, adequate roads), and short of land; but, of course, if we admit all these shortages all we are saying is that we could produce more if the country were bigger all round—which is not a very helpful thing to say.

Both general analysis and the history of the last decade suggest, in fact, that British economic growth is held back mainly by our difficulty in exporting enough. It is true that this can be blamed on a sociological factor, our propensity to a high rate of inflation, or on our strong preference for fixed rates of exchange; but it seems increasingly likely that a major factor is our inability to produce enough manufactured goods of advanced design. The consequent high importance of scientific knowledge is what one would expect, for a densely populated island of meagre natural resources (other than its people) must necessarily import much, and it must pay for these imports by selling goods which embody the skills of its people. These skills and qualities of mind, used to provide superior productive processes or superior products, are indeed our greatest natural resource. Given that we are no longer willing to work harder than the people of other advanced nations, we can only hope to work more effectively. Unless we can embody, in a substantial range of exports of goods (and services), technical qualities or designs which make them desirable in relation to the offerings of other countries, we shall have continually to interrupt our economic progress because of balance of payments crises.

A starting-point for scientific policy, therefore, is the examination of what we know about the pattern of exports ten or twenty years ahead. This is not such a hopeless task as it seems. As the National Economic Development Council has already shown for the next few years, so likewise over a longer period it is certain that the rate of expansion of our exports must be very high if we are to have a reasonable rate of growth. But certain types of export are almost sure to decline, because of known social trends or income changes or political tendencies. For other forms of export the chances of a technical lead appear poor, because our competitors have special advantages of situation or market structure. The remaining trades, whose exports, old or new, stand a chance of selling in increased quantity on their scientific or technological content, are still admittedly numerous; but they can be given a rough ranking according to the urgency of technical development. Attention must also be given to trades which are losing ground in the home market because of increasing imports of superior goods from overseas; for the saving of imports is as desirable as the promotion of exports. For the same reason, research directed to the production of new materials from largely British resources, likely to be preferred on their technical merits to imported materials, would be desirable.

By asking these questions about long-term prospects of British exports (and about substitution for imports) it would, I think, be possible to reach some conclusions, first about applied research and development, then about the pure research which feeds into it, and finally about the forms of training needed to support this research. Thus, a very cursory look will show that the long-run prospects for exports depend on improving the quality of engineering design; and one would have to ask some awkward questions about the rightness of using scarce scientific resources to achieve technical change in industries which (for reasons other than their scientific status) seem certain to decline. By regarding exports as the limiting factor, I have been able to leave out of account the alternative to using our own scientific resources to increase our stock of knowledge, namely, the borrowing of ideas from other countries; for this (though important) is plainly insufficient to give us a technical lead over the original creators of the knowledge.

The use of the limiting factor as a means of deciding a long-range policy can be illustrated in another of its aspects by considering a country much smaller than Britain, and I use as an example the Republic of Ireland. With under three million people and one-fortieth of the British national income, Ireland has a much more acute problem of choice than Britain; a large number of branches of science and technology must inevitably remain unrepresented within her borders. Although she has substantial external assets, some special features of her institutional structure cause her (like Britain) to look much at her export position. The importance of agricultural exports (especially cattle) provides special problems, both of finding markets in a much-regulated industry and of dealing with considerable random and cyclical changes.

A country as small as Ireland must inevitably depend on a flow of ideas from overseas and, in fact, she has attracted numerous branch factories of overseas enterprises, exploiting her advantages of available labour, land and water and of ready entry to the British market. Ireland probably ought to create better facilities for keeping in touch with scientific change on a wide front; for the reason suggested on page 173 above, she cannot afford to specialise as much as her small size would suggest. But she ought also to try to achieve a technical lead, necessarily on a narrow front, so that some part of her exports can sell on quality and not just as a result of the relative cheapness of her labour—an advantage which may disappear, and is in any case offset by high transport costs. An obvious area for achieving a technical advantage is in the processing of agricultural products; it is thus entirely appropriate that, through a state enterprise (the Irish Sugar Company), she is a pioneer in the development of accelerated freeze-drying of foods. Very difficult problems of choice are involved in concentrating sufficient attention on a few subjects in a country of small resources. The influence of the universities tends to produce a spread of research over many units, each too poorly endowed to be effective. Industry is largely organised in small firms with no research of their own. In such a situation, it is particularly important that the state and the nationalised industries should be in a position to make a scientific and careful analysis of the country's needs.

IV

I HAVE so far simplified discussion by assuming that scientists and technologists are needed only for research, development and design; but this, of course, is far from the truth. They are needed also in the control and supervision of production, in the selling of technically advanced products, in general management and administration, and in the training of the next generation. The requirement of scientists for production and sales can to some extent be regarded as derived from their use in research and development, for it often happens that on the adoption of a new product or process those who were responsible for its development move across to manage and control it and sell its output. At least one may expect that the requirements of trained scientists for production and sales departments will increase rapidly, and that it will be related to the previous distribution of research and development effort.

The needs of education and training can at first sight be assessed by a simple exercise in programming; by reducing the numbers of scientists and technologists available for other work now, the rate of growth of numbers in the future can (given appropriate provision of buildings, technicians, etc.) be increased. But there is room for thought also about the efficiency of the educational process. The universities, in particular, are apt to talk as though it is necessary for them to have a large improvement of the staff-student ratio, to train greatly increased numbers, and to use wherever possible graduates with first-class honours. It is not self-evident that a country can rightly ' plough back ' so much of the best products of its educational process, though a justification might be found in the pure research functions of the universities. Certainly it takes little observation of university teaching to suggest that, as a teaching process, it is remarkably inefficient. This means that it is making inefficient use of some particularly scarce resources.

The needs of general management and administration are much harder to assess. Clearly it is important that at certain key points for decision—on boards of directors, among the senior administrators of government departments, and so on—there should be people with some idea of what the relevant science and technology are about. Equally clearly such people are often lacking, and British industry, commerce and administration suffer from the consequent lack of understanding among those who actually make the important decisions. No amount of expert advice can make up for a fundamental lack of ability to understand scientific ideas. But it is difficult to assess how many scientists, and of what kinds, will be absorbed into management and administration; it is perhaps best to leave this as a natural 'leakage' from the other uses of scientific resources, realising that in consequence the numbers available for these other uses will be somewhat reduced.

V

To sum up: I have proposed, as a starting point, an assessment of the economic requirements for science and technology in relation to the most serious limiting factor of the economy concerned. This is an act of long-term planning, and will yield an approximate idea of an appropriate distribution of scientific effort, first for applied research and development, thence for pure research and for the uses of scientists in production, sales, management, administration and education. The plan thus drawn up is only a starting-point, for the country may wish to decide that, at the expense of the flow of material wealth, it will divert resources to satisfy pure intellectual curiosity. It will also need to look at the adequacy of its scientific effort to provide receiving stations for ideas from the rest of the world, and at the general balance between pure and applied research. At many points there will be a temptation to suppose that the problem of the distribution of scientific effort is too difficult, complex or vague to be solved. Yet a few tentative moves towards rationality may be better than leaving the disposal of a valuable resource to random and undirected influences. One hopes that this is why we have in Britain a Minister for Science.

CHOICE AND THE SCIENTIFIC COMMUNITY

I

SINCE the end of the Second World War, public expectations of science have increased enormously. Statesmen, industrialists and writers now look to science for the accomplishment of a great variety of marvels. Frequently there is to be heard the proclamation that science will bring prosperity, either to the world at large, or to some important part of it. All too often, science is applauded as the harbinger of success in the Cold War, much as it used to be thought that victory in the classical armed conflicts of history would belong to the combatants with God on their side. Then again, in the modern world, science is looked to for the cure of cancer and of the other terminal diseases of human beings, so that it often seems to be a kind of fountain-head of immortality. The community has come to regard science as some mysterious amalgam of a philosopher's stone, Holy Grail and monkey gland. The consequences of this development for science itself are necessarily profound.

The first thing to be said is that the great expectations which now mark the relationship between the community and science seem not to be a consequence simply of the steady but rapid growth of the scale of scientific activity in recent decades. After all, the volume of scientific activity has been increasing in an exponential fashion more or less without interruption since the Renaissance, as Price [1] has often pointed out. A century and more ago, industry began to reap a rich harvest of innovation because of the fundamental work of scientists contemporary with Liebig and Faraday. By the end of the century, it had been recognised, at least in some quarters and in some countries, that science could powerfully stimulate industrial progress and, for example, it was in the hope that industrial science could contribute to national strength that the youthful Planck first took up his work with the Berlin company which manufactured electric lamps. Until now, however, the industrial and other benefits of science have most frequently been regarded as accidental and uncovenanted accompaniments of industrial enterprise. It is only in the very recent past that the community as a whole appears to have come to regard science as an instrument by which the character of society might be transformed.

[1] Price, D. J. de S., *Big Science, Little Science* (New York and London: Columbia University Press, 1963).
This article from *Minerva*, II, 2 (Winter, 1964), pp. 141–159.

This development, coupled with the enormous increase there has been in the scale of scientific activity, has tended to restrict the freedom with which scientists decide on the direction of their research. In a great many unaccustomed ways, pressures have been created which are quite extrinsic to science itself but which tend to influence major decisions about the deployment of scientific skill and resources. The feeling that skill in space research is a necessary means of demonstrating national prestige, for example, cannot fail to influence the balance of scientific work in advanced societies as profoundly as did the conviction, a decade ago, that a nation could not be great if it lacked a programme for the exploitation of atomic energy.

If considerations like these have made it harder to be sure that decisions on scientific policy are determined by the long-term interests of science itself, recent developments have also added extra complexity to the making of decisions about the direction of scientific advance. Not merely, for example, is it now necessary for a nation, or for a community of scientists, to decide whether or not it will engage in space research; it is also necessary to decide at what level this will be done. At a rate of £1 million a year, or twice as much? Or even at a rate of £10 million? The latitude with which the financial support for a programme of research may now be determined is so great as to amount not merely to a great quantitative difference between alternative programmes of research but to a qualitative difference as well.

For these and many other reasons, making wise decisions about the spending of effort and resources on scientific research is now more difficult than it has ever been. But the importance of decisions that are wise and accurate has increased with the passage of time. For one thing, there is now much more at stake. The recognition by the community at large that science can be full of promise is a splendid opportunity. Yet the history of science is one long alternation of periods of achievement and periods of stagnation and frustration. Unfortunately there is no guarantee that mere largesse will suffice to ensure that wise decisions will always be made. On the contrary, there is at least a superficial danger that uncritical spending can be an impediment, and not a spur, to the progress of science.

Thus there has grown up an unprecedented need for criteria by which wise decisions can be made about the directions in which scientific efforts should be bent. There is also a great need of mechanisms by which such decisions may be aggregated into effective policies for the administration of what has become a considerable part of the creative energy of modern societies. Superficially it may seem that the demands of pure research on the one hand and of applied science on the other are so different that quite different prescriptions would be necessary to deal with them. This is, however, only an approximation to the truth. In reality, it is not feasible to divide the scientific community into two parts, one of which is concerned with pure science and the other with its application. The two kinds of activity have more in common than casual expectation might suggest.

II

There is one sense in which the making of deliberate choices between alternative courses of action is not merely an adjunct to scientific research but the essence of it. In experimental work good practice entails perceptive planning. A radio-astronomer, for example, does not point his telescope at random to the sky but instead he follows some plan he has chosen for himself and which is intended to provide information bearing, in what is calculated to be a pertinent way, on some problem previously defined. In high-energy physics, experimenters do not use their bubble chambers, spark chambers and other paraphernalia to make indiscriminate observations of the atomic particles in a cyclotron beam. Rather, they choose to select from the multitude of observations which might be made those most likely to provide telling information. Very often the value of an experiment is determined by the way in which the necessary choices are made in the course of its design. Thus there is a point of view from which the elementary creative act in scientific work is the conduct of a choice between alternatives.

The progress of science as a whole can thus be regarded as the product of all the elementary choices made in all fields of science. Since it is unreasonable to expect that all choices will always be the best that could be made, in any particular sense of the word "best", it follows that the development of science as a whole is the resultant of what happens to its separate parts. This happens in much the same way as the behaviour of a macroscopic body (such as a mass of gas) is the resultant of the generally incoherent behaviour of its parts (such as the individual molecules of a gas).

This is entirely in accord with the view of the character of science itself, which has been described by Popper,[2] among others. The everyday work of science consists of the confrontation of hypotheses with data gathered by experiment or observation, or with the implications of other hypotheses. By these means, some hypotheses will be destroyed altogether. Others will, in the course of time, be modified so as not to conflict with what is known from experiment or inferred from theory. But this point of view provides a touchstone for the criticism of scientific work comparable in every respect to

---

[2] Popper, Karl, *The Logic of Scientific Discovery* (London: Hutchinson, 1959) and *Conjectures and Refutations: The Growth of Scientific Knowledge* (London: Routledge and Kegan Paul, 1963).

the criteria of literary merit which have been worked out, over the centuries, by literary critics going back to Aristotle. Briefly, scientific work, experimental or theoretical, is to be valued for the pertinence with which hypotheses are scrutinised and for the way in which the confrontation of a hypothesis by experimental facts or by another hypothesis can suggest the formulation of still other hypotheses.

Thus, in academic science at least, there appears to be no difficulty in specifying the ends that should be sought in the making of individual choices. To say this is not, of course, to describe a mechanism by which sound choices might be made, or even to specify the circumstances in which such a mechanism might most easily be established, but it is a good, if somewhat familiar, starting-point.

It follows immediately, for example, that there is a distinction to be drawn between a scientific work which bears on the known body of understanding and work which, though superficially similar in character, is not directly relevant to the intellectual task of winnowing out hypotheses. Strictly speaking, the undirected gathering of information about natural phenomena is of no immediate scientific value. It would, for example, be quite possible to design and to conduct an experimental programme for recording temperatures at selected sites in the British Isles with an error no greater than, say, 0·0001 degrees Centigrade. To do this would have no scientific value so long as there is no hypothesis to be tested.

The character of scientific discovery also implies that there must be some degree of order in the general advance of scientific understanding. One thing follows from another. It is inconceivable, for example, that Einstein's approximation to the laws of mechanics could have preceded the statement of the Newtonian laws. Indeed, Einstein's hypothesis could not have been lent a semblance of credibility if there had not been Newton's hypothesis to amend. Within narrow fields of science, it may thus be possible to suppose that progress must follow some natural path which can be recognised quite clearly with hindsight even though, almost by definition, it cannot be predicted in advance. The part played by the choices of individual scientists, again only within a sufficiently narrow field, is to determine whether progress along some natural path is steady and sustained or, on the other hand, intermittent and uncertain.

At first sight the natural paths of discovery in different fields of science may seem to be more or less independent of each other but history suggests that there are, in the long term, restrictions on this rule. Repeatedly it has turned out that a rapid advance in one field depends on a technique borrowed from another. How much of biology would there have been if it had not been for the microscope? And how much would there have been of the modern theory of solids if it had not been for the work of the molecular chemists in the thirties, itself dependent on the quantum theories of the previous decade? These antecedents suggest that there is a structure to the pattern of advance in the whole field of science which, though less definite than what seems to be the natural path of advance in a narrow field of science, does imply that there are advantages to be won when different branches of science march forward together and not independently.

As it happens, it is a part of the everyday activity of scientists to predict what connections there are likely to be, in ways such as these, between different branches of science. For example, it is now recognised that it will be possible to form more tangible hypotheses about the nature of the tenuous matter in interplanetary space when there is a better theoretical understanding of the behaviour of tenuous electrified gases or plasmas, in their own right. A better understanding of the interplanetary medium, however, will help to refine hypotheses about the outer envelope of the atmosphere of the sun and about the mechanism by which the sun discharges streams of atomic particles from its interior. But progress in this direction cannot fail to bear significantly on the task of forming a deeper understanding of the energy balance within the atmosphere of the earth.

In ways such as these, science as a whole can be thought to hang together. For the task of making choices between alternative outlets for scientific effort, it is important that a great many of the connections between different fields of activity can be predicted in advance. In other words, it is possible to think of planning what may be described as a balanced strategy for the encouragement of science.

These considerations are important because they imply that, in research directed towards academic ends, some meaning can indeed be given to the notion that some choices are wise and some are not so wise. Evidently this is possible both for the separate decisions, which must be made by individual scientists, and for the much broader issues concerning the general scale and direction of effort in a particular field, which must be made for the scientific community as a whole. To be sure, to know that it is legitimate to seek what may be described as a *best choice* does not imply that it is easy, or even possible in all circumstances, to tell just what the best choice should be. After all, in mathematics it is frequently much easier to prove an existence theorem than to prescribe a technique for calculating the root of an equation, the sum of a series, or some other quantity. For the making of choices about the direction of scientific effort, however, one simple but far-reaching conclusion is immediately apparent: there is no foundation for the view that scientific work which is not planned deliberately in advance may yet be valuable because of what it may accomplish accidentally.

48

The implication of all this is that the character of scientific advance is at once reasonable and coherent. The deepening of understanding is a process of aggregation. Occasionally, it is true, a well-planned experiment, or a daring hypothesis, can send a shaft of light through a whole corner of science, transforming the scene as if in a theatre. But for the most part progress is unspectacular and undramatic and, above all, methodical.

<h2 style="text-align:center">III</h2>

Choosing the applications of science is more difficult than the problems of choice which arise in the academic field, not because technology and science differ in their methods or character but because the tests of success are, in each case, different. In academic science a new hypothesis, or an experiment, or a speculation, can be counted as a success if it leads to a deepening of understanding—to an extension of the bridgehead of certainty which has been established on the continent of ignorance by the last several centuries of modern science. The test, in other words, lies within science itself. These are circumstances in which good judgement can hope to be rewarded for its own sake. In the application of science, on the other hand, success requires not merely that the analysis of a problem and its solution should be executed well but also that the result should be a product or a process which can profitably be made use of commercially.

Evidently choices made in circumstances like these must pay more than passing regard to the commercial considerations. Obviously it would be folly to embark on the development of a new kind of transistor if there were no good reason to believe that such a device would find a market. Equally, there is no point in carrying through the development of a new kind of traction engine if there is no reasonable chance that it will prove to be competitive with existing types of engines. More precisely, a necessary preliminary to the application of science in some field or another is as realistic an assessment as possible of the commercial advantages of successful development. Strictly speaking, this is a task that lies quite outside science itself. Its tools are market research, economic analysis and the accountancy of production engineering. To this extent, it might be supposed that these essentially commercial questions might be separated from the more technical questions of whether some particular development is feasible. Unfortunately, however, reality is more complicated than that.

Science and the application of science are, in many ways, inseparable. For one thing, their practitioners form a single community. An academic scientist may find himself drawn into the application of science. A scientist working in industry may be drawn, or may elect to go, into academic research. By definition, technology or the application of science is dependent on science proper; the opposite is to some small but frequently exaggerated extent also true, for science is frequently seen to depend on technology for its achievements. In this sense it is difficult to know whether raising the issue of whether Thomson would have discovered the electron without the development of improved vacuum pumps is tantamount to asking a question or merely begging a question.

In the public administration of science, however, the most important link between science and its application is that determined by the public expectation of science. For most practical purposes the reason why governments are increasingly prepared to spend large sums of money on the support of science of any kind is that they seek the benefits of technology. Public support for contemporary research in molecular biology, for example, is prompted by the belief, accurate or otherwise, that the result will be better medicine. The fact that the same research is likely to yield epochal transformations of the biological sciences as a whole is not an important part of the public argument for generously supporting this field of science.

In a great many other fields of science it is inevitable that decisions about the deployment of scientific effort cannot be made simply on scientific grounds. It is inevitable that the possible applications of some branch of science should play a part in determining how generously it is supported. Two awkward conclusions can be drawn. For one thing, there may be a genuine conflict between the needs of science and those of technology. The rapid application of some new idea may, for example, require that some gap in basic scientific understanding should be filled in without paying proper regard to the intellectual economy and elegance of scientific inquiry in its own right. Then there is also the real risk that public attention to the possible practical benefits of science may induce a frame of mind in which wishes become the fathers of thoughts and in which a great deal of effort is needlessly expended on work that cannot be profitable, either in scientific or technological terms.

Unfortunately there is now a great deal of evidence that the second of these difficulties has become a great source of confusion in the formation of public policy on science and technology. There is a tendency to believe that everything is possible if only enough is spent on it. This feeling seems to be sustained, in part at least, by examples such as the success with which the Manhattan Project was carried to fruition during the Second World War. That was an occasion on which what seemed to have been an immeasurably great scientific task was successfully completed long before this could have been reasonably expected. It is in the same spirit of unbounded optimism that other technological exploits have since been attempted.

The projects for exploiting thermonuclear fusion in a controlled fashion illustrate how there has been, at various times since the war, a tendency to assume that everything conceivable is also attainable if only enough effort and ingenuity is spent. In reality, of course, expectations like these are likely to be frustrated if they are not based on accurate technical judgement. To be sure, it is sometimes argued that effort spent on projects which fail is not really wasted, for there are uncovenanted benefits in the form of technical training or in the accumulation of scientific information. That, however, is not the point. Effort thus spent must necessarily be less fully rewarded than effort spent on the orderly gathering of scientific understanding.

Thus, in the application of science, an important part of the process of making wise choices of projects to pursue is that of making an accurate judgement of the feasibility of a proposed development. This, as it turns out, is one of the points on which science and technology have most in common. Technologists do not ordinarily commit themselves in advance to the view that some development is at once feasible and commercially profitable, nor do they allow themselves to be identified with prophecies about the time needed to turn a good idea into a reality. The process of development, which may occupy a great deal of time, is strictly analogous to the process in academic science of subjecting some group of hypotheses to test and refinement by experiment. An engineer working with atomic power might, for example, start from the hypothesis that a nuclear reactor based on certain principles is a feasible proposition, and then refine and amend that hypothesis, during the process of development, into the hypothesis that a nuclear reactor with a certain specification could be made a working device, capable of producing power at some calculable cost.

Technology may thus be expected to exhibit a pattern of development resembling in its orderliness the development of a particular narrow field of science. To be sure, it may be necessary, perhaps for commercial reasons, to compress a great many technical developments into a short space of time, so that the logical connections between different parts of a process of development may be thoroughly obscured. This cannot, however, hide the lasting truth that in technology as in pure science, it is possible to make choices from among alternatives in the particular sense that the best choices represent the logically most economical way of carrying through a particular development.

In practice it may easily turn out that a thorough consideration of the circumstances dictates that the best strategy requires that two separate routes to what should be the same goal should be simultaneously pursued.

51

However, this does not modify the principle that, where technical considerations alone are concerned, the application of science in a particular field can be made an orderly process resembling in flavour and in character the orderly advance of science itself.

IV

Of all the choices that must be made in the conduct of academic research, the decisions which must be made by individuals involve the least confusion. If there is an instrument intended to record radio emission from the sun, for example, it is for the experimenters most directly concerned to decide how best to arrange that the measurements will bear, in a significant way, on hypotheses which may exist about the activity of the sun. Should the recordings be made with a device that responds quickly to small changes so that a great deal of information is collected, or, will it suffice (and perhaps be more significant) if the recording device provides an hourly average of the activity of the sun? And, in analysing these recordings, will it be best if a correlation is made with local time on the earth or with the phase of the sun's rotation? These are choices which determine to a large extent the usefulness of scientific work. In the nature of things, they can only be made to best advantage by the people most intimately concerned.

So great, however, is the importance of these decisions, that there is a case for saying that they form as important a part of scientific activity as do the results of an experiment or the conclusions of a theoretical investigation. The truth of this is recognised implicitly in the way in which scientific literature ostensibly pays great attention to the design of experiments and related matters. In reality, however, the importance of this part of the everyday work of science deserves even more attention than it receives. The process of making choices on these elementary matters needs to be regarded not as a matter which is private to an individual experimenter but as an essential part of the proper activity of science.

The value of open, more or less public, discussion of questions such as these has been excellently illustrated by the experience of research institutes such as those for high energy physics and radio-astronomy, at which several investigators or groups of investigators must use one single (and expensive) piece of equipment. Inevitably there is competition for the privilege to use the common equipment. Quite properly, there have to be methods of sharing the time available among the various claimants. Ultimately some body, which is usually a small committee, has to make a more or less judicial decision. In principle, at least, an attempt is made to allocate the use of the commonly shared equipment in such a way that the experiments likely to be most pertinent to the testing of current hypotheses receive the most attention. In this spirit, for example, the committees responsible for the running of the proton synchrotrons at Brookhaven and at CERN (Geneva) appear to have decided, without dissent, that from the beginning of 1963 the greater part of the work done with these gigantic pieces of equipment should be devoted to experiments which bear on hypotheses concerned with the interaction of neutrinos with the other elementary particles of matter.

This does more than merely illustrate the assertion that such research institutes have pioneered the working out of procedures by which scientists may bring their disagreements into the open. For the problem of how to make wise choices in the strategy of science, what matters is that these procedures are at once a tangible embodiment of the principle that the concept of a wise choice is attainable and of how it is practicable to discuss explicitly the question of whether or not a proposed experiment bears in a significant way on a hypothesis to be tested. To be sure, it is by no means always necessary that there should be formal procedures like these for deciding whether individual choices made by working scientists are to be deemed wise or not. Indeed, one of the practical virtues of the way in which universities are constituted so that their members must continually account for their work, and justify it to their colleagues, is precisely that this is a way of arranging that work which is considered most significant should receive the most support, both in physical resources and in the assistance of persons who may be attracted by the intellectual excitement of a well-conceived project.

To emphasise the importance, to the making of wise decisions about the strategy of science, of the intellectual confrontations brought about by the structure of a university is not the triviality that it may seem to be. For one thing, there is every reason to think that such a process of intellectual confrontation will produce decisions which are, in the long run, sounder than those reached by any other method. To be sure, it does not follow that the process can only be carried out within universities. Research institutes and laboratories can provide the same atmosphere. But the essential truth is that there is more reason to trust a choice made by the consensus of interested parties at an outspoken colloquium than one agreed upon by a body of people not equipped to make decisions about such matters. In other words, in science, as in the rest of scholarship, discussion is the most effective means of refining the aggregated stock of understanding.

Though this mechanism can ensure that decisions about the work of individuals are made as wisely as possible, it does not in itself provide a means of ensuring that broader issues are wisely decided. One of the complications of recent years has been the increasing extent to which decisions on the strategy of science have to be made outside the strictly academic community. For one thing, decisions have to be made on a national basis when the scale, or the cost, of an experiment has outgrown the capacity of a single university, or a single laboratory. This is especially true when an expensive or intricate piece of equipment, such as a particle accelerator or radiotelescope, is concerned.

There are also circumstances in which a national decision will be made on the amount of research to be undertaken in a specified field. It was in this spirit that the Advisory Council on Scientific Policy recommended, in the autumn of 1963,[3] that the United Kingdom should not participate in the construction of a new 300,000 Mev accelerator at Geneva until deficiencies in the pattern of research in British universities had been made good. Earlier in 1963, in the United States, the President's Science Advisory Committee had proposed that there should be more financial support for research in oceanography, largely on the grounds that a good deal of the work that could be done with intellectual profit might also turn out to be of great commercial benefit.

On what principles are matters like these to be decided? Here again it is apparent that the essential need is the exercise of scientific judgement similar in every way to the judgement involved in the decisions which must be made on the conduct of individual experiments. This, at least, is the best way for the orderly development of science as a whole. Scientists must make estimates, for themselves and for their colleagues, of the extent to which work in one field or another will be germane to the outstanding tasks of testing and reformulating hypotheses. It is inevitable that they should also consider how far the support of research in some field will serve as a vehicle for education and the extent to which it may be expected to yield practical benefits by application.

The interplay of these arguments is splendidly illustrated in the way in which the *ad hoc* Biological Research Committee of the Royal Society reported in November 1961, on the need for more generous support for the biological sciences. After outlining the ways in which biological research has been transformed in the recent past, the committee argues that " the major consideration concerning biological research in this country is that it should be put in a position to keep up with these advancing and changing frontiers of knowledge ".[4] The committee goes on to argue that

[3] *Annual Report of the Advisory Council on Scientific Policy 1962–1963.* Cmnd. 2163 (London: H.M. Stationery Office, 1963), pp. 4–6.
[4] *Report of the ad hoc Biological Research Committee: A Report presented by the Council of the Royal Society to the Chairman of the Advisory Committee on Scientific Policy in November 1961* reprinted as Appendix B to the *Annual Report of the Advisory Council on Scientific Policy 1961–1962*, Cmnd. 1920 (London: H.M. Stationery Office, 1963), p. 20.

there should be an increase in the scale of university research activity in the biological sciences, partly because it had decided that there would be a great unsatisfied demand for graduates in biology and partly because it estimated that there would be great practical benefits from more biological research.

" One reason for this suggestion is that the two major scientific problems with which mankind is faced—the control of human fertility and the provision (including storage without wastage) of an adequate supply of human foodstuffs—are both biological problems which still call for intense efforts over a very wide field. A second type of reason is that there are very many fields of biology which seem to be nearing the point where they might have practical applications. We have recently seen, for instance, the mass application of insecticides and antibiotics, which, it is not too much to say, has changed the face of the world as a place for human habitation. It is not, in our opinion, unduly optimistic to suggest that there may be several other biological developments of equal practical importance in the offing. To name any particular one of these developments is, of course, speculative but simply to give an indication of the kind of thing we have in mind, might be mentioned the possibility of turning fishing from a hunting into a farming operation, of storing human organs and tissues and being able to utilise them in surgery, of being able to control the growth of tissues so as to facilitate wound healing or even regeneration of lost parts, of being able to eliminate harmful insects by manipulation of the genetic structure of their populations, of much improved control over the breeding of strains of animals and plants to meet particular requirements, of improved methods of utilising photosynthesis for the capture of solar energy, etc." [5]

This statement of the need for more biological research in Britain is important if for no other reason than that it is representative of the ways in which the scientific community seeks to influence development in particular branches of science. Obviously its command of public attention depended to a large extent on the fact that the membership of the committee spanned the whole range of the biological sciences. The carefully qualified suggestion that the investment of public funds in academic biological research would eventually bring practical benefits cannot be thought of as if it were the prospectus of a public company seeking to raise funds in the stock markets. It amounts to nothing more than the assertion that the application of a particular science must wait on the growth of the science itself. As such it is an impeccable statement.

[5.] *Ibid.* p. 22.

In reality, however, this statement and those which resemble it do not reveal the whole of the process by which the scientific community, or an important part of it, has chosen to ask that increased effort should be devoted to biological research. Two issues are especially important. In the first place, it is not apparent from the report of the Biological Research Committee that the consensus of opinion that more biological research is needed in Britain is itself the product of uncounted individual decisions about the appositeness of particular projects for research in the biological sciences. The report of the committee carried weight with the scientific community as well as with the public authorities because, by the autumn of 1961, it had become widely appreciated that the previous decade had thrown up a great many new opportunities for putting old hypotheses to crucial tests and for formulating entirely novel hypotheses about the character of living processes. This tacit recognition by the scientific community of the scientific force of the committee's argument must have been indispensable. On other occasions the pronouncements of committees consisting of equally influential members have not met with such immediate success, largely because the tacit consent of the scientific community was lacking. It follows that the processes of forming scientific opinion about the importance and the potential fruitfulness of some field of research should be brought much more into the open than they are at present.

It is also apparent that the Biological Research Committee was only necessary because the condition of biological research had fallen far behind the true needs of the science, which had been accentuated by the great discoveries in molecular biology at the beginning of the fifties. On the face of things it might appear that the blame for this situation could be laid at the door of the authorities responsible for supporting university research, and it is true that the fifties were years in which university budgets in Britain were far less generous than they should have been. At the same time, however, the universities—and the biologists—were not blameless.

This is a point made by the Advisory Council on Scientific Policy in the report [6] for the year 1961–62 in which it otherwise approved of the work of the Biological Research Committee. Briefly, there is good reason for believing that biologists would have won financial support for their new activities sooner if only they had made their needs more forcefully apparent at an earlier time. It is also true that university departments whose organisation grew up in the thirties and earlier were slow to accept the implications of the new molecular biology, and, in particular, the need to abandon old sub-classifications of biology. That, too, is a defect that might

---

[6] *Annual Report of the Advisory Council on Scientific Policy 1961–1962.* Cmnd. 1920 (London: H.M. Stationery Office, 1963).

have been avoided if only there had been more open discussion of the new scientific opportunities.

So it is that the problems of making decisions about the conduct of individual experiments and about the allocation of scientific effort in different fields of academic research are seen to involve the same kind of judgement—the capacity for making an assessment of the fruitfulness of the confrontation of hypotheses and experiments. It should be no surprise that the freedom and openness of intellectual discussion which is the best guarantee of the soundness of individual research should also be the best means of deciding when, and how, there should be major shifts of emphasis in the general direction of scientific research. Indeed, it would be best for science as a whole if this could be more widely recognised. The scientific community as a whole should more commonly than hitherto make a practice of arguing out in public the reasons which may exist for the choice of some particular course of action or some particular deployment of scientific effort.

Failure to do this can be most damaging. Perhaps the most obvious danger is that a field of research may not be explored as and when the opportunity to do so becomes apparent—the kind of failure to which the Biological Research Committee of the Royal Society drew attention. There is also the danger, however, that from lack of a sufficiently rigorous public discussion, within the scientific community or even within the larger academic community, scientific research will become burdened with a great volume of activity which is not calculated to be in the best interests of the progress of science. It is hard not to believe, for example, that a good deal of the scientific work undertaken in the service of space research is designed primarily to make some use of techniques and instruments which exist, and is not calculated as it should be to put current hypotheses to crucial tests. Not merely does undiscriminating activity of this kind provide a great mass of information which it is hard to assimilate into the body of scientific understanding, but it also tends to interfere with the rigour of the processes of scientific analysis and inquiry upon which, in the last resort, must depend the wisdom of all decisions about the conduct of scientific work.

The implication is that the scientific community should encourage the growth of many more institutions capable of arranging for the confrontation of plans and designs for scientific research. For wise decisions about the direction of individual research, the continued outspokenness of the scientific community is the only lasting assurance. Where major shifts of emphasis in the balance of research are in question, it is essential that recommendations should come from bodies of men who are intimately engaged in the work concerned and who are prepared to base their judgements exclusively on an analysis of the scientific opportunities which novel circumstances may present. Ideally, there would be a great many more *ad hoc* committees like that appointed by the Royal Society to examine the needs of the biological sciences. This is the only way of reaching a balanced judgement on questions such as the importance that should be attached, in a country like Britain, to high-energy nuclear physics, space research, or oceanography. In the nature of things the members of distinguished committees such as the Advisory Council on Scientific Policy are incapable of making these decisions on the basis of their own experience. They must expect to glean advice from other quarters, and especially from the parts of the scientific community which are best placed to form judgement on the potential fruitfulness of actual research.

V

Though circumstances in science and in the application of science may appear to be very different from each other, there is good reason to believe that decisions made about the application of science would also profit from free and open discussion. This, for example, is one of the only sure ways of avoiding the frequent and beguiling danger that some technical development will be pursued long after experience has shown it to be less rewarding than first estimates suggested. In practice, the development laboratories of industrial society abound with development projects whose continued existence is chiefly due to the fact that no decision has been taken to bring them to a halt, or to curtail them. These situations are harmful to science and to the scientific community as a whole, not simply because money may be spent in large amounts without bringing any important reward, but also because scientific skills are thus diverted from more rewarding work.

The devotion of a great part of the world's technical resources to the exploitation of nuclear energy for commercial purposes is a telling illustration of the consequences of suspended technical judgement. A decade ago it appeared to a great many governments that there was great technical benefit to be had from the development of nuclear reactors able to produce nuclear energy from uranium. Various arguments were deployed. Some governments hoped that nuclear power would be cheaper than the conventional fuels upon which they were dependent. Others hoped that nuclear power would make good an absolute shortage of conventional fuel. Still others calculated that the growth of manufacturing industry throughout the world would eventually bring about a world scarcity of conventional fuel, a consequent increase of price, and thus a set of circumstances in which there would be a commercial need of some alternative means of obtaining energy in commercially usable amounts. It is hardly necessary to point out that the more general of these expectations have not been changed by the last decade of experience. It is still calculated, for example, that a time will come at which supplies of fossil fuel will become so much harder to obtain that the price will rise to an uncomfortable extent. For particular governments, however, there have been significant alterations of the relevant circumstances but these have not always been followed by an adjustment of the decision that the development of nuclear power is a matter of the greatest urgency.

This applies, for example, to the development of nuclear power in the United Kingdom. The plan outlined a decade ago, in the White Paper *A Programme for Nuclear Power*,[7] appears to have followed an accurately predicted course. The feasibility of generating nuclear power in bulk, and with safety, has been demonstrated for all to see. Moreover, as the original plan supposed, the cost of building nuclear power stations has decreased steadily with the passage of time. A kW of generating power now costs just half of the price paid in the first nuclear power stations. To be sure, some of the hopes originally expressed that it would be possible economically to build power stations which were technically much more advanced have not yet been fulfilled but no doubt this will happen in the course of time. At the same time, however, there has been a quite unexpected reduction in the cost of building conventional power stations. This has been so great that the first predictions that nuclear power would be competitive with the conventional variety in Britain have not yet been satisfied. The question is what decision should now be made about the character, and the scale, of the nuclear power programme in the United Kingdom?

The first thing to be said is that it is most improbable that the scale and character of the technical development which was judged to be appropriate a decade ago should still be suited to the needs of the future. After all, the rate at which nuclear power stations are to be built will be much lower in the years immediately ahead than those who laid the foundations of the nuclear power programme had imagined. In the circumstances, there is a strong case for believing that a less urgent programme of research and development would be best suited to the circumstances. In other words, there is a case for asking that there should now be a reassessment of the objectives of this programme of development, not because the original programme has been a technical failure, but because it has been undermined by technical developments outside itself. Obviously the work

7 *Nuclear Power: A Programme for Nuclear Power.* Cmd, 9389 (London: H.M. Stationery Office, 1955).

already carried out need not be wasted, for the designs of nuclear power stations already developed can be employed to generate electricity. It does not even follow that fewer people should be engaged on this kind of development. What is necessary is that the pattern of the development which remains should be matched more accurately with the likelihood of nuclear power in Britain becoming commercially competitive not in this decade, and perhaps not even in the next decade, but some time before the end of the century.

These are matters on which technical experts, scientists and technologists alike, are best qualified to provide an authoritative judgement. The methods which must be used are, after all, those necessary in the formation of decisions about the virtues of alternatives in all fields of science, academic or otherwise. But once again, major decisions must be the aggregation of a host of minor decisions about comparatively small issues. In other words, it is not realistic to expect that decisions about the application of science can be made once and for all and then adhered to; it is essential that they should be reviewed continually, and that projects should be abandoned once this seems wise.

It is also plain that it is no easier for persons, or for committees, to make wise decisions about the application of science in which they are not themselves directly engaged than it is for individual scientists, or committees of scientists, to make wise decisions about the direction of scientific research. It follows that the object of good administration in the application of science should be to ensure that technologists actually engaged in the detailed work of applying science to some stated end should, for one thing, be required to work in an intellectual climate in which their efforts are constantly being assessed by colleagues and, in the second place, that decisions about the course of development should be left, as far as possible, to the outcome of such confrontations.

The implications of this principle for the course of technical development in industrial societies are far-reaching. For one thing, it is apparent that the accuracy with which wise judgements can be formed about the course of technical development can be seriously affected by hindering free discussion of technical possibilities. This is, for example, one of the consequences of the kind of secrecy which marks the application of science to defence and in some commercial fields. There is no doubt, for example, that some of the untoward expectations which attended the first steps in the development of controlled thermonuclear fusion were themselves the product of the air of secrecy in which the development had been started.

Restraints on free discussion born of secrecy are more frequently recognised and are possibly less dangerous on that account, than certain other inhibitions commonly to be encountered within or about the scientific community. Frequently, for example, free public discussion is inhibited by the mistaken belief that unfavourable criticism of a piece of scientific work is, at best, likely to be mistaken for disloyalty and, at the worst, may be an invitation of public rejection of science as a whole. Nothing, of course, could be further from the truth. In the long run, society's tolerance of the activities of the scientific community will be determined by the extent to which science appears as an integrated intellectual discipline, concerned to see that choices at all levels are made in the best possible way. It is precisely by allowing to go unchallenged lines of technical development whose promise is unlikely to be fulfilled that the scientific community invites the scorn (and the parsimony) of the rest of society.

In practice all this demands that major decisions about the commitment of large sums of public money to the development of particular scientific ends should be regarded as the responsibility of the scientific community as a whole, so that decisions are not made without the most searching technical inquiry by people intimately engaged on the work concerned. Only in this way will it be possible to avoid the falsification of promises that appear to be implied in the application of science, and the consequent frustration both for the scientific community and for society.

VI

Both the character of science and the experience there has been of its practice in modern society suggest that the wisdom of the choices which need to be made both in academic science and in its application will be determined by the intellectual cohesion of the scientific community as such. For one thing, it is evident that the conduct of scientific work can be, and should be, a reasonable process. One step follows from another in a fashion that makes sense, at least with the benefit of hindsight. It is also plain that the easiest way of being sure that reasonable steps are taken in a sequence that has some of the attributes of being natural is that there should be open intellectual discussion of a course of action and its alternatives. To affirm this is merely, of course, to say again that the intellectual activity of modern science is closely akin to the intellectual activity of scholarship in all its forms but this does not imply that the statement cannot bear repetition. On the contrary, there is good reason to fear that it is now as much in need of repetition as it has ever been.

It also follows that the scientific community must be unafraid of rejecting the over-eager expectations of society at large. There is no doubt

61

that science is one of the instruments by which society will be transformed in the decades ahead, as it has been in the decades past. At the same time, to allow unreal expectations of the potentialities of science to grow up and to remain in society is to give hostages to fortune. It is seriously to be questioned whether the danger of this has been fully appreciated in the recent past. If it had been, would there have been such a great expenditure on the vast scientific-technological enterprise of sending a human pilot to the moon?

Finally, the problems of making wise choices between alternative courses of experiment or theoretical inquiry, or between alternative applications of science, must inevitably colour the relationship between science and the community as it is embodied in the machinery by which governments seek to administer science. The point here is that governments must less frequently regard the scientific community as a kind of tool which can be used for accomplishing one objective or another, but must rather regard the discipline as a field of activity whose benefits must be seized when it is possible, as ripe fruit is picked from a tree. It does not follow that the social benefits of science cannot be foreseen in advance. Governments can hope to administer science effectively and in such a way that society wins the greatest return for its investment of money and effort. But the intimate relationship which exists between the making of choices between scientific alternatives and the intellectual liveliness of the scientific community as a whole inescapably implies that a robust connection between the two must be established. Creating and maintaining that connection is the responsibility as much of the scientific community as of interested governments.

THE COMPLEXITY OF SCIENTIFIC CHOICE:
A STOCKTAKING

STEPHEN TOULMIN

I

THE questions about selection and priorities implicit in all discussions of science policy are both difficult and inescapable. They are *inescapable* for the less-developed and the industrialised countries alike: for the former, because large-scale patronage of researches irrelevant to their urgent social and economic needs is a luxury they can scarcely afford,[1] and for the latter because, once calls for expenditure on research and development exceed some 1½ to 2½ per cent. of the gross national product, the sums involved inevitably—and properly—become subject to political scrutiny alongside other comparable items in the national budget, such as social insurance, schools and defence. (Evidently, this point has now been reached even in the U.S.A.: witness the current cut-back in the overseas operations of the National Institutes of Health and the difficult passage to which Congress exposed the latest appropriations for the National Science Foundation.) Questions about scientific priorities are *difficult* in two distinct ways. They are difficult, first, because we are still sheerly ignorant about many of the relevant factors and relationships—*e.g.*, what long-term repercussions certain fundamental but expensive lines of research, such as neutrino physics, will eventually have on the rest of pure science, or how the results of scientific work can be most effectively translated into industrial innovations. (Part of this ignorance we can do nothing about, except wait for new insights within natural science itself: part we could remedy in time, given a systematic study of the sociological, economic and organisational questions involved.) But questions about priorities are difficult also for a second reason: because the essential problems are still to some extent *out of focus*. Though we can see, and state, many of the considerations having a significant bearing on these problems, we cannot yet put them all together so that they yield a coherent and consistent administrative doctrine. This second class of difficulties we should be able to do something about at once, since the problems they pose are essentially analytical ones. What is required in this case is not new knowledge but a clearer and crisper vision of the questions

[1] *Cf.* Dedijer, S., " Underdeveloped Science in Underdeveloped Countries ", *Minerva*, II (Autumn, 1963), 1, pp. 61–81; and also Carter, C. F., " The Distribution of Scientific Effort ", *Minerva*, I (Winter. 1963). 2, especially pp. 173–174, 179–180.
This article from *Minerva*, II, 3 (Spring, 1964), pp. 343–359.

actually at issue in the formulation and administration of a science policy; and our first aim must therefore be to remove any fog due to ambiguities, cross-purposes or hidden assumptions.

For present purposes, I shall confine myself to this last, limited task. This will be, so to say, a stocktaking. When Edward Shils stated the aims of *Minerva* in his editorial manifesto, he declared that, " by the improvement of understanding, it [*Minerva*] hopes to make scientific and academic policy more reasonable and realistic ",[2] and already the periodical has published four important articles on the government of science: *viz.*, Michael Polanyi's " The Republic of Science: Its Political and Economic Theory ",[3] Alvin M. Weinberg's " Criteria for Scientific Choice ",[4] C. F. Carter's " The Distribution of Scientific Effort ",[5] and John Maddox's " Choice and the Scientific Community ".[6] The subject has meanwhile been very near the surface throughout the public debate about the Trend recommendations [7] and some powerful sidelights have been thrown on it in a collection of essays recently issued by the Council for Atomic Age Studies of Columbia University, New York.[8] So this is perhaps an appropriate time to review the *Minerva* debate about the central problems of scientific choice. Here I shall concentrate on three questions: (i) Is there any consistency of theme or approach in the four articles on this topic published to date? (ii) Where the different authors appear at cross-purposes, can we do anything—*e.g.*, draw any necessary distinctions—to resolve these conflicts, and so put our problems in truer proportion? (iii) If so, are these distinctions already reflected in administrative practice and in the structure of our scientific institutions, or can some corresponding element of cross-purposes be recognised in current procedures also?

II

To start with, we must pick out from the four articles in question the key ideas put forward about scientific choice—even at the cost of simplifying rich and suggestive arguments. (i) Polanyi's central doctrine is that " the body of scientists, as a whole ", constitutes a " Society of Explorers ", which he names " the Republic of Science ".

> Such a society strives towards an unknown future, which it believes to be accessible and worth achieving. In the case of scientists, the explorers

---

[2] *Minerva*, I (Autumn, 1962), 1, p. 5.
[3] *Minerva*, I (Autumn, 1962), 1, pp. 54–73.
[4] *Minerva*, I (Winter, 1963), 2, pp. 159–171.
[5] *Minerva*, I (Winter, 1963), 2, pp. 172–181.
[6] *Minerva*, II (Winter, 1964), 2, pp. 141–159.
[7] See, *e.g.*, Rt. Hon. Aubrey Jones, M.P., " Some Comments on Trend ", in *The Technologist*, I (January, 1964), 1, pp. 19–24, referred to below.
[8] Gilpin, Robert, & Wright, Christopher (ed.), *Scientists and National Policy-Making* (New York and London: Columbia University Press, 1964).

strive towards a hidden reality, for the sake of intellectual satisfaction. And as they satisfy themselves, they enlighten all men and are thus helping society to fulfil its obligation towards intellectual self-improvement.[9]

Internally, the scientific community has a structure by which its intellectual values are continually cross-checked and maintained. Each scientist is " a member of a group of overlapping competences ", while " the whole of science " is " covered by chains and networks of overlapping neighbourhoods ". As a result,

> Scientific opinion is an opinion not held by any single human mind, but one which, split into thousands of fragments, is held by a multitude of individuals, each of whom endorses the other's opinion at second hand, by relying on the consensual chains which link him to all the others through a sequence of overlapping neighbourhoods.[10]

Questions about " scientific merit ", Polanyi argues, must be judged by the tribunal of working scientists forming the relevant " neighbourhood ". Science can accordingly advance only if the republic of science is left free to determine research priorities by what one may perhaps call its own " syndicalist " procedures. Only so can " this vast domain of collective creativity " be " effectively promoted and coordinated ".[11] Despite " the generous sentiments which actuate the aspiration of guiding the progress of science into socially beneficent channels ", this aspiration has been proved, in actual experience, to be " impossible and nonsensical ". Indeed, " any attempt at guiding scientific research towards a purpose other than its own is an attempt to deflect it from the advancement of science "—and a self-frustrating attempt at that, since science " can advance only by essentially unpredictable steps, pursuing problems of its own, and the practical benefits of these advances will be incidental and hence doubly unpredictable ".[12]

(ii) Maddox agrees with Polanyi that " decisions on scientific policy " should be " determined by the long-term interests of science itself ",[13] but he adds some interesting points. First, he rejects the idea that science must inevitably be fragmented into small, overlapping sub-disciplines:

> Antecedents suggest that there is a structure to the pattern of advance in the whole field of science which, though less definite than what seems to be the natural path of advance in a narrow field of science, does imply that there are advantages to be won when different branches of science march forward together and not independently . . . In other words, it is possible to think of planning what may be described as a balanced strategy for the encouragement of science.[14]

[9] *Minerva*, I (Autumn, 1962), 1, p. 72.
[10] *Ibid.*, pp. 59–60.
[11] *Ibid.*, p. 61.
[12] *Ibid.*, p. 62.
[13] II (Winter, 1964), 2, p. 142.
[14] *Ibid.*, p. 145.

Secondly, he sees that Polanyi's analysis has left one major question unanswered: *viz.*, how we are to extend into the technological field the kind of debate by which research priorities are argued out within pure science. Evidently, " it is not feasible to divide the scientific community into two parts, one of which is concerned with pure science and the other with its application " [15]: yet how—bearing in mind the commercial considerations which in this case are unavoidable—are decisions about technological merit to be arrived at? Maddox recognises only one sure method. Wise decisions, in technology as in pure science, depend on " intellectual confrontations " like those by which pure scientists argue out the merits of their projects within the academic world [16]:

> Ideally, there would be a great many more *ad hoc* committees like that appointed by the Royal Society to examine the needs of the biological sciences. This is *the only way* [my italics, S.E.T.] of reaching a balanced judgement on questions such as the importance that should be attached, in a country like Britain, to high-energy nuclear physics, space research, or oceanography . . . It is precisely by allowing to go unchallenged lines of technical development whose promise is unlikely to be fulfilled that the scientific community invites the scorn (and the parsimony) of the rest of society.[17]

True, Maddox concedes, the Royal Society's *ad hoc* Biological Research Committee is not an entirely happy precedent. It did its work 10 years too late, and even now the universities are " slow to accept the implications of the new molecular biology "; but he sees no alternative—" That, too, is a defect that might have been avoided if only there had been more open discussions of the new scientific opportunities." [18]

(iii) Weinberg adds some important refinements to a similar general view. The difficult choices (he sees) are not those within any single branch of science, but those which pit different fields of work against each other:

> The five fields I choose [for comparison] are molecular biology, high-energy physics, nuclear energy, manned-space exploration, and the behavioural sciences . . . [These fields] are incommensurable: how can one measure the merit of behavioural sciences and nuclear energy on the same scale of values? Yet the choices between scientific fields will eventually have to be made whether we like it or not. Criteria for scientific choice will be most useful only if they can be applied to seemingly incommensurable situations.[19]

To justify these invidious decisions in the public domain, Weinberg proposes three groups of criteria, which he refers to as " technological merit ", " scientific merit " and " social merit " respectively. Technological merit is an uncontroversial idea: it represents the normal balance between

[15] *Ibid.*, p. 143.
[16] *Ibid.*, p. 150.
[17] *Ibid.*, pp. 155, 158.
[18] *Ibid.*, pp. 153–154.
[19] I (Winter, 1963), p. 167.

research costs and prospective return with which the directors of all science-based industries are familiar. Scientific merit, for Weinberg, is to be measured as much by indirect repercussions as by direct promise: " that field has the most scientific merit which contributes most heavily to and illuminates most brightly its neighbouring scientific disciplines ".[20] Questions of social merit, however, thrust scientific choice into the arena of politics and human values: they have to do with such things as health, food production, defence and prestige. To deserve really massive public support, a research field should rate highly on more than one of these three scales. Molecular biology (in his opinion) has all the merits, high-energy physics is currently somewhat overrated, while space-research is only masquerading as science: " If we do space-research because of prestige, then we should ask whether we get more prestige from a man on the moon than from successful control of the waterlogging problem in Pakistan's Indus Valley Basin. If we do space-research because of its military implications, we ought to say so. . ." [21] To sum up:

> As much out of prudent concern for their own survival as for any loftier motive, . . . scientists must acquire the habit of scrutinising what they do from a broader point of view than has been their custom . ... For scientists as a class to imply that science can, at this stage in human development, be made the main business of humanity is irresponsible—and, from the scientist's point of view, highly dangerous.[22]

(iv) By this point, we have already moved some way from Polanyi's uncompromising " syndicalism "; but with Carter's paper we plunge into another world. For his approach is deliberately utilitarian: " The aggregate scientific effort of a country is a function of the number of scientists and technologists (of various kinds) which it contains, and of the effectiveness of their use . . . I will assume that it is national policy to have full employment of its scientists, and to use their skill as fully as our present understanding of the ' use of skill ' allows." [23] Any nation is of course at liberty—as a matter of deliberate policy—" to undertake pure research beyond what can be given a broad economic justification by its ultimate application ", and it would so demonstrate that it " regarded intellectual discovery as more important than an increment of wealth. . . . In point of fact, however, it would be difficult to prove from the British record that the total scale of pure research exceeds what can be given an economic justification." [24] So Carter makes the economic criterion his starting point:

> Though this is not entirely an economic matter, it is only from economics that any guidance will at present be obtained. In other words, it is possible

[20] *Ibid.*, p. 166 (Italicised in original).
[21] *Ibid.*, p. 170.
[22] *Ibid.*, p. 171.
[23] *Ibid.*, p. 174.
[24] *Ibid.*, p. 175.

to give some sort of an answer to the question: What kind of distribution of scientific effort will most effectively increase the flow of wealth? [25]

Applying general economic principles to the present situation of Britain (as contrasted with—say—China), Carter concludes that the country is moving inexorably into a " Swiss " or " Irish " position:

> Unless we can embody, in a substantial range of exports of goods (and services), technical qualities or designs which make them desirable . . . we shall have continually to interrupt our economic progress because of balance of payments crises.
>
> A starting-point for scientific policy, therefore, is the examination of what we know about the pattern of exports ten or twenty years ahead.[26]

How will such an examination work out in practice? " By asking these questions . . . it would, I think, be possible to reach some conclusions, first about applied research and development, then about the pure research which feeds into it, and finally about the forms of training needed to support this research." [27]

The general moral of Carter's discussion is thus the reverse of Polanyi's: whereas Polanyi roundly declares that " any authority which would undertake to direct the work of the scientist centrally would bring the progress of science virtually to a standstill ",[28] Carter subordinates major decisions about scientific policy to a glorified N.E.D.C. or *Commissariat-Général du Plan*:

> This is an act of long-term planning . . . The plan thus drawn up is only a starting-point, for the country may wish to decide that, at the expense of the flow of material wealth, it will divert resources to satisfy pure intellectual curiosity . . . Yet a few tentative moves towards rationality may be better than leaving the disposal of a valuable resource to random and undirected influences. One hopes that this is why we have in Britain a Minister for Science.[29]

III

Thus my first question answers itself. In his approach to questions about scientific policy, an economist like Carter shares almost no common ground with a scientist like Polanyi. For him, the fundamental significance of scientific research lies in its long-term contribution to the stock of productive knowledge: the " intellectual curiosity " of the natural scientist is an object of disinterested patronage alone, a sideline into which—if the nation so decides—surplus resources may be diverted " at the expense of the flow of material wealth " From this standpoint, Polanyi's conception of a self-governing republic of science, " helping society towards intellectual

[25] *Ibid.*
[26] *Ibid.*, p. 178.
[27] *Ibid.*, p. 179.
[28] I (Autumn 1962), 1, p. 56.
[29] I (Winter 1963), 2, p. 181.

self-improvement ", must appear little more than a visionary ideal, out of touch with economic realities. From Polanyi's point of view, in its turn, Carter's own account must at first glance look like a piece of short-sighted materialism, which threatens to destroy the social fabric of science, and so bring it to a halt. Admittedly, Weinberg and Maddox go a little further than Polanyi towards conceding the claims of technological merit and commercial demand; but all three scientists would—it is clear—prefer to see research priorities based very largely on " scientific merit " and the technological dividends of scientific discovery treated as uncovenanted mercies.

We have, then, an economist's view, according to which science is basically deserving of support because it is the handmaid of industrial growth; and a scientist's view, representing technology as a kind of scientific roulette in which those who plunge deepest tend to win the biggest prizes. Taking either view at its face value, one could perhaps grade scientific pro-jects in an " order of priorities "; but the resulting orders of merit—and the science policies based on them—would presumably be quite different. What, if anything, can be done to moderate the apparent opposition between them?

The crucial step, I believe, is to recognise that *there is no single problem of scientific choice*. Questions about research priorities arise for govern-ment (to say nothing of industry) at many different points and on many different levels and there is inevitably, as a result, *a plurality both of " criteria of choice " and of " orders of merit "*. Unless some particular administrative context is specified, questions about " scientific choice " become essentially indeterminate, for it is the exigencies of the specific context which impose particular criteria of choice. (The National Economic Development Council and the Medical Research Council adopt, naturally and properly, different standpoints and different standards.) Conversely, it is the very demand for a single *overall* " order of merit " that lands discussions of science policy in confusion. When members of the Advisory Council on Scientific Policy ask themselves whether they should recommend allocating the next £10 million to C.E.R.N., to cancer research or to space studies, they are understandably flummoxed. But this (I shall argue) is not their fault. In a well-ordered administration, problems that involved " measuring the merit of incommensurable . . . scientific fields . . . on the same scale of values " [30] ought not to arise; and if they do, this is a failure of government. For the " cardinal choices " [31] in national policy are—not

---

[30] *Cf.* Weinberg, A. M., " Criteria for Scientific Choice ", *Minerva*, I (Winter 1963), 2, p. 167.

[31] This phrase, introduced by C. P. Snow in his Godkin Lectures—*Science and Government* (Harvard University Press, 1961), p. 1—is in course of being transformed into a useful " term of art ": see, *e.g.*, the Columbia symposium, *Scientists and National Policy-Making* (footnote 8 above), especially the essay by Albert Wohlstetter, pp. 174 *et seq.*

surprisingly—*political* ones, to be made at cabinet (or presidential) level; and, by the time they reach that exalted status, the points for decision will have had to be restated, so ceasing to be technical questions about scientific importance, and becoming political questions about the rival claims of health, defence, intellectual patronage, higher education and the rest.

IV

Let me argue this case concisely, and so in broad terms. (a) First, I shall criticise the false unity suggested by Maddox's phrase " the scientific community " and by Polanyi's corresponding phrase " the body of scientists, as a whole ": these exaggerate the contrast, both in qualifications and in interests, between " scientists " and " the rest ", and so help to conceal the plurality of ways in which " research " (to use a less loaded word than " science ") impinges on the national life. (b) Next, I shall contrast a few of the different ways, and administrative contexts, in which questions about research priorities and choices arise: in some contexts (as we shall see) there is great virtue in Polanyi's insistence that the republic of science must be autonomous, but in others any claim that " scientists " should be the sole judges is unsustainable. (c) Finally, I shall take up some questions about administration and organisation which arise at the points where technical issues abut on, and shade into, political ones.

How, then, are we to define " the scientific community "? There are two separate difficulties. (i) This " community " has highly blurred edges: it is (as Professor Wallace S. Sayre puts it in the Columbia symposium) " a world of uncertain boundaries "—

> Who are the members of the scientific community? Is it an open community, hospitable to all who desire to enter, or is it open only to those who meet severe tests of eligibility? More specifically, are there " hard scientists ", whose membership is taken for granted, and " soft scientists ", whose credentials are dubious? Are physicists and chemists members of the scientific community by right, while other natural scientists must submit additional claims for admission? Do all engineers qualify, or only certain types of engineers? Do doctors of medicine qualify, or only research scientists in medicine? Are social scientists full members of the scientific community? . . .[32]

Sayre himself hints that the phrase is inevitably tendentious—

> [It] may thus belong in that class of invocations, so familiar to the political process, which summon up numbers and legitimacy for a point of view by asserting that " the American people ", or " the public ", or " all informed observers ", or " the experts " demand this or that. There is nothing especially astonishing about this, since all participants in the political process indulge in the stratagem . . . but there may be grounds

[32] Gilpin, Robert, & Wright, Christopher (ed.), *op. cit.*, p. 98.

for mild surprise that the code of science permits its extensive use by scientists either as deliberate strategy or in genuine innocence.[33]

Elsewhere in the symposium, the size of the American " scientific community " is estimated—using different indices—at anything from 1,200,000 (the totality of professional groups) down to 200 (the effective political lobby).[34] Could we do any better in Britain? I think not. For the same quandaries apply here. We need not question the " good standing " as scientists of Fellows of the Royal Society—leaving aside honorary fellows like Sir Winston Churchill; but it is notorious that " the Royal " pays too little attention to technology,[35] while positively excluding workers in the human, economic and social sciences, even from those branches which have long since earned their admission into the scientific academies of other countries.

(ii) Further, " the scientific community "—however it is defined—has an internal structure which can sometimes be of great significance: notably, an age-and-status structure. This creates practical difficulties, which Polanyi and Maddox both underestimate, when it comes to assessing " scientific opinion ". The republic of science is not, in practice, a full democracy: in its external affairs especially, it is a *gerontocracy,* and this fact causes me to wonder whether " open intellectual confrontations " and " many more *ad hoc* [Royal Society] committees " really represent (as Maddox declares) " the only way " of reaching balanced judgements about scientific priorities. It should be illuminating, for instance, to experiment with the " Delphi Method ",[36] developed by the Rand Corporation to deal with just such problems. In this technique, a panel of specialists participates in a sequence of questionnaires, through which they compare and contrast their own opinions against those of other panel-members, *without knowing their identities*: so a consensus of opinions can be achieved which is undistorted by considerations of seniority and status. Suppose, for example, separate " profiles " of expectations and priorities had been established in this way in 1952, among biologists aged from 25 to 35, and among biologists aged from 50 up. Might this not have helped to short-circuit or accelerate the procedures of the Royal Society Committee on Biological Research? Might not the cardinal importance of molecular biology have, as a result, been publicly established rather earlier than 1960?

---

[33] *Ibid.,* pp. 98–99.

[34] *Cf., ibid.* the essays by Robert C. Wood and Christopher Wright, especially pp. 48, 273.

[35] *Cf.,* Sir Gordon Sutherland's proposal for a separate Royal Society for technology on the lines of the Swedish *Ingeniörs Vetenskaps Akademien,* reprinted from the *Guardian* in *The Technologist,* I (January 1964), 1, pp. 39–40.

[36] See, *e.g.,* Dalkey, Norman, & Helmer, Olaf, " An Experimental Application of the Delphi Method to the Use of Experts ", *Rand Corporation Memorandum,* RM-727-PR (abridged), July 1962.

V

The "scientific community" (I imply) cannot be isolated, as a single, coherent body of opinions and interests, clearly separable from those of, say, hospital nurses, architects and town-planners, industrial managers and citizens-at-large. Rather, we must begin by classifying the different ways in which research can enter into and affect the national life, as a preparation for understanding the variety of ways in which such research can interact with and depend on government. For if science is inevitably brought into contact with government at a variety of points and in a variety of ways, then " scientists " can have no *unique* standpoint *vis-à-vis* government: there will be many legitimate modes of interaction, in each of which scientists apply their minds to a different group of problems, and the needs of each such partnership will impose their own pattern of research priorities and criteria of choice.

Evidently enough, one can distinguish at least four distinct types of research: (i) pure research in the natural sciences, engaged in with no eye to utility or material productivity; (ii) speculative technology, designed to broaden the general base of man's practical capacities; (iii) product-oriented research, aimed at developing a material, machine or medicine, say, to meet a given specification; and (iv) problem-oriented research, aimed similarly at finding an acceptable solution to some practical problem in the field of, say, transport, public health or defence. Different government agencies have correspondingly varied reasons for being interested in research and for underwriting its costs. It is a general interest of any country, for instance, (i) that its intellectual and artistic institutions should maintain a healthy level of activity, through either public or private patronage. (In Britain at present, the fiscal structure provides little incentive for the " patronage of the intellect " by corporations or individuals —the " seven-year-covenant " rule is an obstacle unknown in, say, the U.S.A.—so this function falls largely on government.) Similarly, government has proper parts to play, by (ii) helping to broaden the country's base of technological skills, as an essential element in guaranteeing economic development, (iii) paying the cost of developing certain types of product— *e.g.*, aircraft—and (iv) underwriting research on ways of overcoming the major problems of modern communal life: the Buchanan inquiry into the future shape of towns [37] is a classic example of this last class.

All this may seem very obvious, if stated baldly; but it is directly relevant to the *Minerva* debate on scientific choice. For some of the contributions to that debate manifestly concentrate too exclusively on

[37] *Traffic in Towns. A Study of the Long Term Problems of Traffic in Urban Areas. Reports of the Steering Group and Working Group appointed by the Minister of Transport* (London: H.M. Stationery Office, 1963).

one particular relation between science and government. Suppose, for instance, that all government support for research properly fell under the heading of " patronage of the intellect " (category (i) above), then indeed Polanyi's doctrine might be the whole answer. Certainly, within the limits of this category, the intellectual promise of a scientific project and the significance of its outcome are matters, above all, for the scientific guild. Even there, history suggests qualifications: a few exceptional scientific discoveries have had repercussions spreading indirectly to other intellectual disciplines, and beyond—Darwin's theory of natural selection is a clear illustration. To concede Polanyi's essential point, however, it is no more the business of government to pronounce on these wider implications than on the narrower scientific merits of an inquiry. They are a matter for the broader academic confederation, of which the republic of science represents, so to speak, one " canton ".

Again, Carter's criteria of choice become relevant only when we leave category (i) behind and move into categories (ii) and (iii). (Does not Carter explicitly set aside " the satisfaction of pure intellectual curiosity " [38] as irrelevant to his discussion?) From the point of view of the National Economic Development Council, the crucial question is, indeed, into which sectors of technology Britain—with its special position and problems— should plunge most deeply, as promising the highest economic return on investment in research and development. In this discussion, the scientific guild can again make an indispensable contribution, since judgements of *feasibility* are an integral part of the task; but scientists cannot have the entire say, since the final criteria of choice must here be based on a balance of *economic* considerations, in addition to intellectual ones. As for categories (iii) and (iv): there the " customer " too must have a say, along with the scientist and the economist. Criteria of merit in pharma- ceutical development, for instance, involve—or *should* involve—something more than estimates of feasibility and financial return: the interests of the citizen-at-large (*qua* potential patient) must be explicitly represented also. Finally, research into a social problem like the impact of road traffic on city life calls for wider representation still. The fruits of research in this case will be far more than merely intellectual: what is at issue is the whole environment in which our children and grandchildren will live. In the post-Buchanan debate, accordingly, architects, town-planners and citizens-at-large are all entitled to intervene; and the creation of Professor Buchanan's Department of Transport at Imperial College indicates that—at last—the solution of *social* problems is becoming recognised in this country as being a species of science or technology, as

[38] *Minerva*, I (Winter, 1963), 2, p. 181.

truly as the refinement of quantum electrodynamics or the development of thermoplastics.

So, in real life, the republic of science cannot stand apart from the general commonwealth. Back in the 1930s, Polanyi's campaign to defend the autonomy of science against projects for a Nosey-Parkerish state centralism had a real point. By the 1960s, the need for academic science to be self-governing seems to be being conceded even in Russia and Polanyi's protestations are—surely—more insistent than they need be. As the social sciences too approach their coming-of-age, his distinction between the republic of science and the rest of the community becomes excessively disjunctive. The urgent question to-day is, rather, how the self-governing republic of science is to be integrated, not only into the broader academic confederation, but into the whole community of citizens. For it is on the answer to this question that our broader criteria of scientific choice ultimately depend.

VI

If carried only so far, this analysis may read like an exercise in pure philosophy. But, if one further premise is taken into account, it begins to have practical implications in the field of public administration. For there is a general principle of organisation which holds in the administration of scientific affairs as forcibly as it does in the rest of the public service. This is the Chalk-and-Cheese Principle: namely, that the structure of departments and advisory committees should be so ordered that, at each point, decisions have to be taken *between commensurable alternatives*. In a well-structured administration, choice always involves, say, signing a contract for one fighter-bomber rather than another fighter-bomber, allowing a pension to one ex-serviceman rather than another ex-serviceman, laying down one set of guide-rules for tax exemption rather than another alternative set of guide-rules—comparing, in each case, chalk with chalk and cheese with cheese. Administration can, in fact, be efficient and equitable only if its organisation is—in this sense—*functional*: when departments or committees are regularly required to choose between chalk and cheese, their decisions are inevitably more difficult and more political.[39]

In most areas of public administration this general principle is well recognised, and organisational structures have been developed which

---

[39] Let me digress at this point. It is often questioned whether an undergraduate training in Oxford philosophy is a useful preparation for the civil service higher administrative grades; yet the young assistant principals who took part in a recent discussion which I led at the Treasury Centre for Administrative Studies found it entirely natural to adapt the techniques of analytical philosophy to the discussion of public administration. What I here call respecting the Chalk-and-Cheese Principle, for instance, they called—following Gilbert Ryle—" avoiding category-mistakes ", and the sophistication of Oxford philosophical debate served them well in the discussion.

faithfully reflect, and discriminate, the different governmental functions to be performed. Yet there is reason to think that, in the case of scientific affairs, such an equilibrium has not yet been reached. This can be illustrated by referring, briefly, to the Atomic Energy Authority in Britain and to the National Aeronautics and Space Administration in the U.S.A. To begin with the A.E.A.: from the outset its functions have been ambiguous and the activities of its stations have contributed to the nation's power, health, defence and pure science programmes—to mention only the most obvious applications. In the first wave of enthusiasm for atomic technology, it was understandable that the A.E.A. should be set up as a unitary agency with a single central budget and parliamentary vote. But the time seems now to have come when government-sponsored atomic research might alternatively be budgeted for through a number of different channels, according to function. All these activities pre-empt large funds and highly-qualified staff, which might be being employed in other directions instead. It is, accordingly, arguable that the pure physics done under the A.E.A. wing stand—first and foremost—to be weighed in a scale against the pure physics done at universities, the atomic power programme against other work on the generation of power, the work on medical isotopes against other medical research, etc. For each of these sectors of work must, in the last instance, be judged *politically* as elements in the overall governmental outlay on pure science, fuel and power, etc.

Similar comments apply to N.A.S.A. On the one hand, N.A.S.A. initiates and supports a complex of activities which impinge on the national life of the U.S.A. in a great variety of ways: it plays a part in weapons development, engages in meteorological surveys, finances research both in universities and in its own scientific laboratories, fosters much more-or-less speculative technology, and maintains the newly-renamed " aerospace " industry in the manner of life to which it wishes to remain accustomed. N.A.S.A.'s functions, that is to say, are *multiple*. Yet it is funded by a *unitary* appropriation and, despite Congressional scrutiny, the final decision how this fund shall be divided between the different functions is—inevitably—taken within N.A.S.A. itself. In this way, N.A.S.A. has been given both the authority and the means to alter substantially the national division of effort as between (say) industrial development, defence and scientific research. Instead of being a purely administrative agency, it is to that extent a political agency taking political decisions: a " state within the state ", to which Congress has delegated some of its powers under the Constitution.

To say this is not, of course, to *attack* N.A.S.A. The point at issue is not the manner in which the agency has in fact exercised these delegated powers: it is the constitutional question, whether such a delegation of political authority by Congress to a largely autonomous agency best serves the national interest. Or would it be better if the multiplicity of N.A.S.A.'s functions were paralleled by a multiplicity in the sources of its funds? Is there not a case, as with the British A.E.A., for balancing each of N.A.S.A.'s major programmes against similar programmes maintained elsewhere from other sources as parts of the overall federal effort in the fields cf (say) defence or technology or pure science?

Corresponding ambiguities arise, in Britain, over the present status of the Advisory Council on Scientific Policy. Members of the council speak with feeling about the " incommensurable alternatives " between which they are required to choose. Should they, for instance, recommend spending more on particle physics at the expense of cancer research, or *vice versa*? The Chalk-and-Cheese Principle prompts one to ask whether the A.C.S.P. is in fact being *asked the right questions*. True: a phrase like " cancer research " may cover anything from clinical trials to basic cell-physiology and this kind of ambiguity helps to conceal the problem. For the A.C.S.P. may quite fairly be asked to say whether, as a matter of pure intellectual patronage, there is a fair balance in the scale of government support as between (*e.g.*) particle physics and basic cell-physiology. But if " cancer research " is taken in a wider sense, it becomes a genuinely *medical* matter which should be weighed, not against the government's support for pure science, but as part of its overall provision for expenditure on health. So understood, the choice between particle physics and cancer research becomes a decision whether to allocate more funds (a) to the patronage of the intellect or (b) to improving the nation's health. *This is not a technical choice, but a political one.* As such, it should be taken by the cabinet on its own responsibility, not referred to a scientific committee like the A.C.S.P. The most any advisory council can do in this case is to appraise the likelihood of getting tangible results in the cancer field by a given technique and for a given outlay—after that, the choice is up to the government itself.

In this respect, the positions in the Washington hierarchy of the President's Science Advisory Committee and the Office of Science and Technology are less ambiguous. Instead of being attached to one of many executive departments, P.S.A.C. and O.S.T. are White House bodies, responsible directly to the Chief Executive and his Cabinet.[40] They are

---

40 There has been some sharp criticism recently of the science advisory machinery in Washington—see, for instance, Abelson, P. H., " A Critical Appraisal of Government Research Policy ", *Robert A. Welch Foundation Research Bulletin No. 14*: Houston, Texas (November, 1963). However, in considering this debate, one must take care to distinguish questions about the personal position of Dr. Jerome Wiesner in the Kennedy administration from questions about the institutional merits and defects of

accordingly in a position to answer the technical questions bearing on all issues of national policy, whether these fall within the area of defence, industrial development, transport, education, intellectual patronage, or anything else; and their advice serves, not to *supplant* the ultimate political decisions, but to make them *better-informed*.

Unfortunately, the British A.C.S.P. has been answerable, not to the Cabinet, but to the Lord President of the Council in his role as Minister for Science. Its brief has been to watch over the functioning of the research councils and grant-giving agencies through which the government exercises its function as patron of scientific research, so as to ensure that " science-as-a-whole " is getting the right kind of support. Yet, on the argument of the present paper, the idea of " science-as-a-whole " has the same defects as the idea of " the scientific community ": one can no more separate off " science " from the rest of the national life than one can separate " scientists " from the rest of the population. More and more, the contributions of research to the welfare of the community are overlapping into spheres of the national life far removed from the pure patronage of intellectual inquiry. This fact is *both* the justification for the vast increase in public funds being devoted to research today *and* the reason why the older systems of organisation and budgeting—dating back to the Haldane Report of 1918 [41]—are breaking down. Research on health, housing, transport, criminology and the rest can today be regarded as integral parts of the nation's overall activities in the fields of medicine, housing, transport, crime-prevention, etc.: the pre-1939 conception of the Lord President's Office, as the principal channel of patronage for research of all kinds, has outlived its usefulness.

VII

Though I started from a very different point, I end up in a position very close to that expressed in a recent article by Aubrey Jones:

> The Haldane Committee was perfectly right to point to the desirability of research both within administrative departments and outside them. I incline to the view, however, that the Trend Report exaggerated when it says that the Haldane Committee laid down a principle about the doing of research outside administrative departments. I question whether there is any principle involved. The doing of research by external agencies is a

the P.S.A.C. system itself. Evidently, the longstanding and extremely close personal relationship betwen Wiesner and John F. Kennedy gave rise among other scientists to tensions and suspicions resembling those provoked in Britain by the somewhat similar relationship between Churchill and Cherwell. It is interesting, as a result, to compare Dr. Abelson's attack on Wiesner's position with C. P. Snow's famous polemic against Cherwell, in his Godkin lectures on *Science and Government* (Harvard University Press, 1961).

[41] *Report of the Machinery of Government Committee.* Chairman: Lord Haldane. Cd. 9230 (London: H.M. Stationery Office, 1918).

convenient administrative device for certain purposes at certain moments of time. But the greater the use of external agencies the more certain it is that the research will be distant from the purposes of the departments affected. . . I believe that some forms of research, and notably industrial research, should be moved closer to the departments affected.[42]

Still, this statement of the problem perhaps needs to be qualified. The quandary left partly unresolved by the Trend Committee [43] is, whether research laboratories wholly supported by government (such as the Road Research Laboratory) should be " under " the Department of Scientific and Industrial Research or " under " the department most closely affected (the Ministry of Transport, say). British traditions being what they are, it is commonly taken for granted that the department on whose parliamentary vote such a laboratory is carried must be given administrative control over it: " Who pays the piper calls the tune." Yet here again Britain may perhaps learn from American experience. For this administrative quandary reflects a novel conflict of considerations, of a kind which is likely to become more frequent rather than less. In the organisation of scientific research, the natural unit is the *technique*—nuclear reactors, rockets, lasers, computers, games-theory or whatever. One and the same technique may well contribute to the national life in half-a-dozen ways and so interest several government departments. The examples of the A.E.A. and N.A.S.A. posed the question, whether this unity of technique requires a unitary *budget*: in the present case the question is, rather, whether a unitary budget requires *administrative* integration into the user-department. Must a fully-supported research organisation always be administratively " under " some particular ministry?

The problem of combining scientific techniques, administrative flexibility and financial responsibility is solved in a new way in the " research contract " system, which has been exploited far more fully in the U.S.A. than in Britain. Research organisations such as the Rand Corporation and the Stanford Research Institute could scarcely survive without government finance; yet administratively they are independent entities, free to develop the techniques around which their activities revolve as they see fit, and able to " sell " their services to different agencies within government proper. While preserving the unity of technique which was the original *raison d'être* of the A.E.A. and N.A.S.A., they differ from those agencies in drawing their funds from a plurality of sources.[44] Finance reflects function,

[42] Jones, Rt. Hon. Aubrey, P.C., M.P., " Some Comments on Trend ", *The Technologist*, I (January 1964), 1, p. 24.

[43] *Committee of Enquiry into the Organisation of Civil Science.* Chairman: Sir Burke Trend. Cmnd. 2171. (London: H.M.S.O., 1963).

[44] On the development of the research contract system, and related project and grant systems, much fascinating material is to be found in the recent Kistiakowsky Report: *Federal Support of Basic Research in Institutions of Higher Learning* (Washington: National Academy of Sciences, 1964).

organisation reflects technique: while serving many departments, they are integral parts of none. This pattern has, up to now, been followed in Britain scarcely at all.

No doubt the research contract system would need some modification before it could successfully be transplanted into the British environment. Yet, at the very least, its merits and shortcomings deserve to be considered with some care. For this is one of the many new models of organisation which must be borne in mind, as the administration of Britain progressively adapts itself to the task of supervising all those complex and manifold functions which research—physical, biological, technological and social research alike—performs in a modern community.

CRITERIA FOR SCIENTIFIC CHOICE II:
THE TWO CULTURES

ALVIN M. WEINBERG

The Financial Support of " Science as a Whole "

IN a previous paper [1] I proposed criteria which could be invoked in judging how to allocate support to different, competing branches of basic science. Such allocations seem to be necessary because what society is willing to spend on all of science is not enough to satisfy every worthy claim on the total funds available for science. I turn now to the broader question: what criteria can society use in deciding how much it can allocate to science as a whole rather than to competing activities such as education, social security, foreign aid and the like?

That such a question can assume any urgency is in itself remarkable. To have suggested that the Federal Government of the United States would be spending about 3 per cent. of the gross national product for research and development would have been unbelievable 25 years ago. Most of the new attitude toward government support of science and technology was prompted by war and fear of war. In the mind of the public, scientific strength has been equated with military strength. Support of science at first was only dimly distinguished from support of the military. But this attitude is changing, partly because the thermonuclear stalemate seems to have reduced our fear of war, partly because the fantastic successes of modern science have begun to penetrate the awareness of the public. Science *per se*, as a valid human activity supported by the public, has acquired some standing, possibly analogous to that of religion in the era before the separation of church and state. As science has become big, it has acquired imperatives, just like any other activity of government, to expand and to demand an increasing share of public resources, and now, for the first time, it has become big enough to compete seriously for money with other major activities of government.

The criteria for choice between different fields of basic science I proposed earlier were of two kinds—internal and external. Internal criteria could be established entirely within the scientific field being considered; these criteria arise from the question: how competently is this field of science performed? External criteria could be established only from outside and

---

[1] " Criteria for Scientific Choice ", *Minerva*, I (Winter, 1963), 2, pp. 159–171. This article from *Minerva*. III, 1 (Autumn, 1964), pp. 3–14.

answered the questions: does this field of science illuminate other fields of science; does it further desirable technological goals; does it further broad social goals? My main point was that a good rating according to the internal criteria was a necessary but not sufficient condition for large-scale public support of a field of science. Only if a field rated highly according to criteria generated outside its own universe could it properly expect large-scale support by society.

In so far as the support of science as a whole can be viewed as different from support of each of the separate branches and kinds of science, I believe one can apply analogous criteria. Society, in its support of science, assumes that science is a competent, responsible undertaking. But society is justified in asking more than this of " science as a whole ". However vaguely stated, society expects science somehow to serve certain social goals outside science itself. It applies criteria from without science—broadly, criteria concerned with human values—when it assesses the proper role of " science as a whole " relative to other activities. We scientists concede this implicitly when we agree that responsibility for choosing between science and other activities belongs primarily to the non-scientists —the members of legislative bodies or the head of the executive branch of government and his staff. In the language of Stephen Toulmin [2] the choice between " science done for its own sake " and other activities of the society is a political choice, as contrasted to an administrative choice, and it is to be made by politicians.

The ordering of human values upon which such choices must ultimately be based is a philosophical question into which I will not enter here. I shall assume that we have decided on social goals and shall then ask how we can translate these into practical recipes for deciding how much science we can afford.

The Budgetary Separation of Pure and Applied Science

I shall dispose of the question of what fraction of society's overall effort should go into " science as a whole " by arguing, along with many others, that " science as a whole " is a misleading idea. The basis for the claim which applied science makes on society is so different from that of pure science that lumping them together clouds the issue. Pure and applied science ought not to be viewed as competing for money.

Applied science is done to achieve certain ends which usually lie outside of science. When we decide how much we should allocate to a project in applied science, we at least implicitly assess whether we can achieve the

---

2 " The Complexity of Scientific Choice: A Stocktaking ", *Minerva*, II (Spring, 1964), 3, pp. 343–359.

particular end better by scientific research than by some other means. For example, suppose we wish to control the growth of population in India and suppose we have at our disposal 200×10^6 per year for this purpose. We could devote most of this sum to investigating fertility, to developing better contraceptive techniques, or to studying relevant social structures in some Indian village. Or, alternatively, we could use the money to buy and distribute existing contraceptive equipment, such as Gräffenberg rings, perhaps using some of the money as incentive payment to induce women to accept the technique. Which way we spend our money is a matter of tactics; evidently no general proposition can tell us how much of our effort ought to be spent on research rather than on practice in trying to achieve effective birth control in India. The scientific work that goes toward solving this problem ought to compete for money with alternative, non-scientific means of controlling the growth of population in India rather than with the study of, say, the genetic code. More generally, where a piece of research is done to further an end which society has identified as desirable, support for this type of scientific work should be considered as part of the bill for achieving the end, not as part of the "science budget". Only that scientific research which is pursued to further an end arising or lying within science itself should be included in our "science budget".

This view has become quite popular in many recent discussions of the subject.[3] It is appealing to the scientist because setting support for applied science outside the science budget reduces the latter enormously—from 16×10^9 to perhaps 1×10^9. At this level the whole question of choice between scientific and non-scientific activities becomes much less significant.

But this stratagem is not as clearly justified as it appears at first sight. Ruling applied science to be part of the budget of non-scientific activities, not of the scientific budget, does not eliminate competition between applied science and basic science. Applied science requires at a secondary level, by and large, the same kind of people as does basic science. Building a large accelerator engages electrical engineers who would otherwise be available to help design control systems for rockets. In allocating support for a given applied science, one must keep in mind the effect of such allocation on basic science, and in supporting basic science, one must keep in mind the effect on applied science. Edward Teller has argued that because of the great emphasis on basic sciences in our universities, we have created an atmosphere that is uncongenial to applied science. He insists that our important applied scientific undertakings suffer because we tend to direct our best talents to basic science, our not-quite best to applied

---

[3] A particularly cogent presentation of this position is made by Stephen Toulmin, *op. cit.*

science. Though Teller's contention is difficult to prove, my own experience supports his view.

A second difficulty is that the aim of any given branch of applied science tends to become diffuse as time goes on. The scientific work of any of the large " mission-oriented " government agencies started out specifically to further the mission of the agency. But as time has passed, these clearly defined, " mission-oriented " goals of applied scientific work have become fuzzy. Byways that originally were germane to the mission flourish—an investigation that began as a promising approach to solve an applied problem, 10 years later becomes an interesting study pursued for its own sake, yet it continues to be described as " applied science ". Thus to leave applied science out of the science budget would leave out a large amount of research which was at one time motivated by an extra-scientific or applied end, but which is now pursued primarily because it is scientifically interesting to those carrying on the research.

Finally, the motivation for basic science is itself often less than pure. Is nuclear structure physics done to further science or to help build reactors? Is the structure of natural products pursued as a challenge to scientific virtuosity in organic chemistry or because out of such studies will come the knowledge of enzyme action which ultimately will lead to control of metabolic disorders? Thus consideration of support of basic research completely apart from applied research is not as clearly defined a proposition as many proponents of this position hold.

Nevertheless, I believe the *general* principle of not considering the budget for applied research as part of our " national science budget " and including only basic research in it has one overriding advantage. By allocating funds to any applied research as a certain fraction of the budget of the (usually non-scientific) activity to which the research is intended to contribute, we keep straight our reasons for supporting the applied research. What this fraction should be must depend on internal criteria—such as, do we see ways of making progress, or are good research workers available? It probably should also depend on the impact that support of that field will have on neighbouring basic fields.

The fraction of effort that goes into achievement of a broad end—like aid to underdeveloped countries or national defence—by scientific research instead of by non-scientific action can hardly be decided entirely by the scientists. The scientific approach to solutions of difficult social problems is becoming increasingly popular. Yet in at least some proposals for action of which I am aware—notably in foreign aid and control of world population—it seems to me that excessive claims were made for science. Scientists alone, when asked to judge how to solve a complex social problem, more often than not recommend more science—just as high-energy physicists, when asked to recommend a programme in basic science, will ask for more high-energy physicists or oceanographers for more oceanography. To overstate the capacities of scientific research as a technique for settling difficult social questions is no more sensible than it is to understate them. Thus, just as I have argued that scientific panels, judging how much money should be allocated to one branch of science rather than to another, should include representatives of neighbouring branches of science, so a panel determining how much scientific research rather than "engineering" or "production" will best achieve a certain non-scientific end should include non-scientists as well as scientists.

Support for Basic Science as a Branch of High Culture

I have argued in the foregoing that applied and basic science should have separate budgets and that the budget for applied science should be set as a certain fraction of the effort allocated to the end (usually non-scientific) which applied science furthers. To this extent I have avoided the problem of choice between " science as a whole " and other human activities by denying the usefulness of the concept " science as a whole ". This still leaves the question of basic science—the science which cannot be justified by any reason except that it satisfies human curiosity. Are there some broad social ends, outside of basic science, which basic science serves, and to which its budget can be tied?

Obviously, some parts of basic science are important to applied science: in my view a much larger fraction of basic science is germane to applied science than many of my basic scientific colleagues are willing to concede. The bulk of the biological sciences is, in a sense, applied. For example, the most recondite and ingenious elucidation of the genetic map of *E. coli* is germane to the whole question of genetic abnormalities. (I often find it amusing to argue with my biologist friends that most of what they do is applied research—that the important distinction in a field of science intrinsically so close to human affairs as is biology is not between " applied " and " basic " but between " intelligent " and " unintelligent ".) Or again, plasma physics, a purely basic science, is central to thermonuclear research, an applied science which is pursued because we wish to enlarge mankind's energy resources.

It is natural to propose that such basic research receive a certain fraction of the resources going into the applied research which it underlies. Every good applied research laboratory allocates to basic research a certain fraction of the resources allocated to it for its related applied research. The ratio of basic to applied research often is very high and is usually highest in the applied laboratories which have had the most success in accomplishing their technological mission. What I suggest is that on the national scale, also, basic research be considered as a fixed charge on the applied research effort, wherever the basic research is intended to contribute to a field of applied science. In making an assessment of relevance, I would incline toward a broad interpretation: for example, I would consider the case of most research in biology as a proper overhead charge to be assessed against the resources allocated to agricultural and medical research.

But what about those fields of basic research, a few of them very expensive, which are really very remote from any applied scientific problems, which are pursued primarily because the researchers find the science intensely interesting, often because the findings in this field are likely to illuminate neighbouring branches of basic science? To what can we tie the allocation of effort for such activities?

This is the most puzzling of all the questions concerning public support of science and any proposed solution must be put forward most tentatively. For basic science of this kind is primarily a somewhat disinterested intellectual activity, in the same sense as are music, literature and art. Indeed, the analogy between the creative arts and this purest kind of basic science is sufficiently great to suggest that, insofar as it must make the choice, society might choose between the pure basic sciences on the one hand and the creative arts on the other. In allocating support for the purest basic research, our allocations for the other creative activities of man might be taken as our guide.

There are many analogies between the purest basic research activity and artistic activity. Each is an intensely individual experience the effect of which transcends itself. The product of each is immortal—the theory of relativity, just as surely as *Hamlet* or the Mona Lisa. Each is concerned with truth—the highest of human manifestations—the one with scientific truth (which deals with the regularities in human experience), the other with artistic truth (which deals with the individuality of human experience).[4] Each enriches our life in unmeasurable though highly significant ways. Each belongs not only to its creator or discoverer, but to all mankind.

In a competition for support between pure science and the arts, I see two major arguments—one that supports the claim of science and the other, the claim of art. The argument that favours science (aside from the obvious one, to which I shall return, that even the remotest pure science may eventually have practical application) is that scientific truth, being based on what we observe in nature, is publicly verifiable, whereas artistic truth,

---

4 This point was illuminated for me in Barzun, Jacques, *Science: The Glorious Entertainment* (New York: Harper & Row, 1964), p. 227 *et seq.*

not subject to the same kind of control, is not publicly verifiable. Artistic critics disagree just as often as they agree. They have no objective and impartial arbiter, nature, to say what is true and what is not true. The truth of science, on the other hand, is rigorously and publicly tested by experiment or by observation or, in the case of mathematics, by logic. Scientific criticism weeds out scientific nonsense more efficiently than artistic criticism weeds out artistic nonsense because, ultimately, science is monitored by a universal and approachable critic, nature, whereas art has no comparable critic. Scientific research and thought, in their mutual and ruthless criticism which reach ever more strongly towards a whole consistent structure, are embedded in what Michael Polanyi has called the " Republic of Science " [5]—the entire scientific community whose mutual interaction is governed by rules of scientific conduct that are themselves laid down by nature, the great scientific lawgiver. The republic of science forces science to be a responsible undertaking, at least in the sense that what science does is true and, in some approximation, true forever. The corresponding republic of the arts has no such final arbiter that can force art to be as responsible as science. In so far as public support ought to go for the more responsible undertaking, the purest science in this regard merits more support than do the arts.

But there is another argument which at present favours the arts. Pure science—that is, science which does not have foreseeable practical applications, such as elementary particle physics or cosmology—is by and large an arcane enterprise which is appreciated mainly by its practitioners. The arts, on the other hand, are generally less restricted in their audience: many more people in the world today can gain enjoyment from listening to Beethoven's *Ninth Symphony* than they can from reading Schroedinger's paper on quantisation as an *eigen*-value problem. Granted that the intellectual delight experienced by the creator in pure science matches that of the creator in art, the direct products of the latter's efforts at present probably give more enjoyment to more people than do the products of the former. Of course, in so far as even the purest science may eventually result in practical applications, it too affects the public at large; but we are speaking here of the science whose practical application is minimal.

The well-paid pure scientists among my friends will undoubtedly object to being converted into scientific bohemians shivering in poorly heated garrets. But I don't think pure science is doomed to that poor an existence if our society decides, even now, to support it on about the same scale as it supports the arts. It is true that the arts are supported poorly by

[5] Polanyi, M., " The Republic of Science: Its Political and Economic Theory ", *Minerva*, I (Autumn, 1962), 1, pp. 54–73.

government, but the total paid by the society, *i.e.*, private individuals and associations, governments, local and federal, for the arts is not negligible and the support is growing. In estimating the total support we give to the arts, we must include the value of theatre admissions, the value of books, better magazines and good records, the total that goes to our performing arts, as well as the direct subsidies in the form of grants to creative artists. The total spent by the United States on all activities that one way or another are concerned with the arts amounted in 1960 to around $2,500 million.[6] Only a fraction of this amount is spent directly by the federal government but this is not relevant. Pure science, unlike music or literature, produces no directly saleable commodity and so if it is to be supported at all by the public it must be supported by the public through its government.

Moreover, it seems likely, with the increase in leisure, and the decrease in the amount we spend on armaments, that a larger and larger fraction of our national income will go into the arts. Voices have been raised favouring a National Arts Foundation, paralleling the National Science Foundation. To make of pure science an avenue for expression of our creative intellectual energy, quite parallel to August Heckscher's [7] proposal to make of the arts such an instrument, strikes me as highly appealing. This latter viewpoint was stated eloquently by N. N. Semenov,[8] the Soviet chemist; he visualises science in the world of the future being appreciated and practised as widely as are the arts in the world today—every man a scientist, to the extent of his intellectual capacity.

I put forward the idea that the purest science be supported in the same spirit and at roughly the same level as the arts as only one among several possibilities. The arts, after all, are not the only non-scientific activity which gives deep intellectual or spiritual satisfaction. For example, religion even today gives great spiritual satisfaction to many people—in our country to many more than do the arts or sciences. And indeed, a case can be made for using the level of support of religion instead of art as a yardstick for how much pure science our society ought to support.

And yet, despite the analogies between science and art, or between science and religion, the idea of relating the degree of support of one to the degree of support of the other is somehow forced and artificial and not really satisfactory. In the long run how much our society is going to

6 Estimate by A. Mitchell of Stanford Research Institute, as reported by *Business Week*, 19 Januar , 1963, p. 68.

7 Heckscher, August, " The Arts in the 1980s ", a lecture at Oswego State University, Oswego, New York (1964).

8 Semenov, N. N., " The World of the Future ", *The Bulletin of the Atomic Scientists*, XX (February, 1964), pp. 10–15; the same idea was also expressed by George Bernard Shaw in *Back to Methuselah*.

spend on basic science depends upon the extent to which non-scientists develop the intellectual power and taste to appreciate, if not to discover, science. The question then is: is it really likely that society will develop so congenial an attitude towards science—say as congenial an attitude as it now displays towards the arts or religion—that it will support the basic scientist at the level he thinks he needs?

Most scientists believe that society will be missing something very important should it *not* develop such an attitude towards pure science. Every scientist knows that much of the satisfaction he derives from his scientific career comes not only from his own original discoveries, but also from the thrill he experiences when he understands, for the first time, someone else's great discovery. My own experience during the past half dozen years illustrates the point. During these years, at least five major discoveries have been made in physics: the Mössbauer effect, the overthrow of parity, the laser, the superconducting magnet, and the SU_3 symmetry in strong interactions. After each of these discoveries I blessed my decision to study physics, since only because I knew some physics could I experience the unique intellectual satisfaction that appreciation of a discovery, almost as much as the discovery itself, affords.

One need not be a great intellect to appreciate a scientific discovery, at least enough to give one real satisfaction. I would guess that all those intelligent enough to take a university degree could learn enough to appreciate some branch of science: if not the most sophisticated parts, then at least the simpler parts. Nor is it necessary for all the public to understand all of basic science. Just as science itself has fragmented under the pressure of the information explosion, so I visualise that " lay-scientists " would also form somewhat separate communities: perhaps there might develop the equivalent of " molecular biology fan clubs ", " high-energy fan clubs ", " oceanography fan clubs ", even as we now have amateur astronomers, radio " hams " and hi-fi enthusiasts.

To educate so many people to a point where they can achieve a sense of participation in the march of science poses a major problem. The scientists themselves will have to spend much effort conveying their message, in intelligible terms, to the rest of society. They will have to deal sympathetically (much more so than I think they do now) with the scientific popularisers and with the scientific educators. If the scientists and their para-scientific associates are unable to convey this sense of scientific adventure to the community that supports them, I cannot see how the purest basic research can, in the long run, expect to receive the support it will demand in the future.

The problem faced by the future scientists has been stated by Professor Eugene Wigner as follows:

> . . . we all hope that the present competitions for the most powerful military posture will become unnecessary soon—perhaps in 10 years, perhaps in 20 years. Quite likely, not only will the present unquestioning support of science cease then; it will be replaced by distrust and even unpopularity. Nobody likes his companions and helpers of a past life from which he has turned to a better one. What will be the role of science then, where the scientist will be no longer a source of power of the government, after having been pampered so long, is not entirely pleasant to contemplate. However, it may be useful. Science that is useless in the sense that it does not help to satisfy other cravings, is still one of the noblest endeavours of man; it would be most pitiful if mankind turned away from science just when it will have the leisure to pursue science in its more noble form.[9]

Support for Basic Science as an Overhead Charge on Applied Science and Technology

I confess to a residual scepticism about our society acquiring this sophistication in the short run, which means, for the working scientist, the years until his retirement. It is probably utopian—as much as Shaw's *Back to Methuselah*—to expect every man in the street to become an amateur scientist or even a science fan.

Thus, much as I hope that our society will acquire this scientific sophistication, it seems clear that in the near, as opposed to the distant, future we shall have to present a more realistic claim on society's support of basic science done for the sheer intellectual pleasure it affords its practitioners. I therefore return to my earlier suggestion that basic science in fields clearly relevant to applied science be viewed as an overhead charge on that particular applied science—that is, against the political mission the applied science is intended to accomplish. I would extend the idea and urge that the purest basic science be viewed as an overhead charge on the society's entire technical enterprise—a burden that is assessed on the whole activity because, in a general and indirect sort of way, such basic science is expected eventually to contribute to the technological system as a whole. In some cases, the help will turn out to be direct, as when a discovery in cosmology illuminates a point in nuclear structure physics; in more cases it will be indirect as when a professor, whose research is in an abstruse field of mathematics, inspires a young engineering student with the beauties of the classical calculus of variations.

Some such view of the relation of the purest basic science to the entire technical enterprise was implicit in Executive Order 10521 issued in

---

9 " Prospects in Nuclear Science ", an address delivered at the Twentieth Anniversary Celebration, Oak Ridge National Laboratory, 4 November, 1963.

1954 by President Eisenhower concerning the terms of reference of the newly founded National Science Foundation:

> As now or hereafter authorised or permitted by law, the foundation shall be increasingly responsible for providing support by the federal government for general-purpose basic research through contracts and grants. The conduct and support by other federal agencies of basic research in areas which are closely related to their missions is recognised as important and desirable, especially in response to current national needs, and shall continue.

From this point of view one has further reduced the dimension of the problem of how much " science " shall we support. Applied science and engineering have already been ruled to be outside the " science as a whole " budget, inasmuch as they are a means of achieving a politically defined mission. Basic science, which is closely related to an applied science (such as biology, *vis-à-vis* medicine), is an overhead assessed against the related applied science, and therefore its level of support is again tied closely to a politically defined end. And finally, the purest basic science, viewed as an overhead against the entire enterprise, would, in analogous fashion, receive support at a level determined as a fraction of the entire remaining technical enterprise. What this fraction should be would itself be a political decision—but if all such research is supported by a National Science Foundation, as suggested by the Executive Order of President Eisenhower, this political decision would amount each year to setting the budget of the National Science Foundation. Of course this political decision would be influenced in part by the public's attitude towards science; but it would also be influenced by the attitude of legislators who are probably more inclined towards science than is the general public, since so much of the business of national legislative bodies now involves science and engineering in one way or another.

Where do the criteria of choice I proposed in my previous paper fit into such a scheme? As I see the matter now, they would be used both by mission-oriented agencies in making administrative decisions with respect to different kinds of basic science and by a body with very broad terms of reference, independently of any technological task such as those given to the National Science Foundation in the United States, in choosing between different basic fields. Within each allocation of funds made for a politically defined task there will always be more claimants than there are funds and choices will still have to be made. The beauty of the idea of basic research as a " scientific overhead " is that it reduces the size of each allocation of funds for scientific research to a more manageable proportion.

Thus I have turned a full circle: I began by asking how much " science as a whole " our society could afford. In developing my views, I have successively reduced the magnitude of science which competes with society's other activities, first by ruling the costs of applied science to be overhead charges on the tasks it sought to further; secondly by ruling the costs of mission-related basic science to be an overhead charge on mission-related applied science; and now by suggesting that the purest science be an overhead on the entire technological system. This is not to say that I object to the view of " science as culture ", a view which places science *per se* directly in competition with other activities of the society. It is merely that, in the short term, basic science viewed as an overhead charge on technology is a more practical way of justifying basic science than is basic science viewed as an analogue of art. Until and unless our society acquires the sophistication needed to appreciate basic science adequately, we can hardly expect to find in the admittedly lofty view of " science as culture " a basis for support at the level which we scientists believe to be proper and in the best interests both of society and of the scientists.

RESEARCH AND ECONOMIC GROWTH—
WHAT SHOULD WE EXPECT?

B. R. WILLIAMS

THE use of science in agriculture, industry and medicine has made possible enormous increases in population, material standards of living, health and the expectation of life. We can expect further increases—provided that we use science for economic growth and not for nuclear destruction.

It is widely believed that the key factor in this growth is the rate of expenditure on research and development. It is also widely believed that the proportion of national output devoted to research and development is critical. Think how often it is argued in Britain that growth is held down by a failure to spend on research and development as high a percentage of national product as do the Americans. In France (and Germany and Australia) the argument tends to be that growth is held down by a failure to spend as high a percentage as the British. Recently, a leading member of the French planning commission assured me that finding the appropriate level of expenditure on research in France was really very easy: "for where expenditure is below the corresponding British level you know that it should be increased ".

There is, however, no obvious logical step from the observed effects of applied science on past growth to the conclusion that *national* expenditure on *research* and *development* is the key to future national growth. Research is simply a process of adding to scientific knowledge. Sometimes new scientific knowledge has a direct influence on the technology embodied in production processes. Frequently, however, a further (and often very expensive) activity called development is required before science can affect technology. Hence the need to distinguish between science—the sum total of systematic and formulated knowledge about the real world—and technology—the sum total of formulated knowledge of the industrial arts. It is technology, and the efficient use of it, that is critical in growth. What evidence is there that in this sense the use of science is a certain derivative of current or recent research and development expenditure?

Evidence of Growth as a Function of Research and Development

In a paper on " The Economics of Research and Development " [1] Dr. J. R. Minasion set out to test the hypothesis " that productivity increases

---

[1] Minasion, J. R., " The Economics of Research and Development ", in *The Rate and Direction of Inventive Activity*, Special Conference Series No. 13 (Princeton: National Bureau of Economic Research, 1962), pp. 93–142.
This article from *Minerva*, III, 1 (Autumn, 1964), pp. 57–71.

are associated with investment in the improvement of technology, and the greater the expenditures for research and development the greater the rate of growth of productivity ". Minasion used a cross-section study of 18 firms in the chemical industry for the years 1947–57. He found that " the relevant research and development expenditure was a highly significant

DIAGRAM I [2]

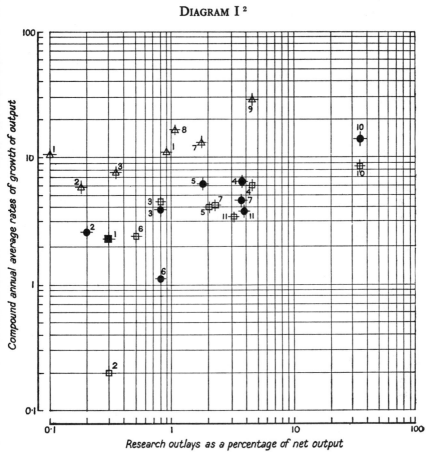

Research outlays as a percentage of net output

● United States
△ Hungary
□ United Kingdom

Key

1	Food	7	Machinery
2	Textiles	8	Electrical engineering
3	Paper	9	Precision engineering
4	Chemicals	10	Aircraft
5	Rubber	11	Total industry (excluding mining, electrical energy and aircraft)
6	Metallurgy		

[2] From U.N. report on *Some Factors in Economic Growth in Europe During the 1950s* (Geneva : United Nations, 1964), Chapter 5, p. 10.

independent variable explaining not only the rate of growth in productivity but also the trend of the profitability of 18 chemical firms in the sample ". He also found that lagged research and development explained end-of-period profitability better than end-of-period research, and development was explained by lagged profitability and that there was not a statistically significant relationship between the growth in productivity and the rate of change in output, average plant size, or the rate of growth in plant size. However, Minasion's sample was not necessarily representative of the chemical industry and he was unable to isolate factors such as royalty payments, which would almost certainly have influenced productivity changes. Nor, of course, could these results be legitimately generalised for the whole economy.

If we shift the measure from growth of productivity to growth of output, we do, however, find some broad but striking relationships between growth in output and research outlays as a percentage of net output. In Diagram I this relationship is plotted for 10 industry groups in the U.K., U.S.A. and Hungary for the years 1949–59.

Dr. R. A. Ewell had earlier pointed to this relationship between research and growth in U.S. industry from 1928–53.[3] He also pointed to a high correlation between the growth of national product and the country's expenditure on research and development. Assuming a causal link between research expenditure and growth he forecast that, if (on 1954 prices) research and development expenditures were increased to $6·3–6·9 thousand million by 1964, gross national product (G.N.P.) would rise to $500 thousand million. Roughly the Ewell projections were:

TABLE I

	1954	1960	1964
G.N.P. index	100	114	131
R. & D. as a percentage of G.N.P.	1·1	1·23	1·37

Ewell's confidence in these conclusions was increased by an alternative line of calculation—that between 1928 and 1953 new products created by research and development were between 11 and 22 per cent. of G.N.P. and that without them annual growth would not have been 3 per cent. but 2½–2 per cent. By imputing this growth to the median research and development expenditure for the period, he concluded that the annual yield to expenditure on research and development was 100–200 per cent. In this calculation of

---

[3] Ewell, R. A., " First outpost in a new frontier ", *Chemical & Engineering News*, 18 July, 1955.

yield Ewell made no allowance for outlays other than those on research and development, although on his calculations $11 of capital expenditure were required for $1 of research and development.

Implications of This and Other Evidence

Minasion's conclusions were based on a very limited sample—18 U.S. chemical firms for the years 1947–57. In Britain for the period 1949–59 no such conclusions emerged from the Freeman and Evely analysis of a sample of 44 firms in general engineering, 22 in chemicals, 12 in electrical engineering and 17 in steel. For the sample firms in chemicals and general engineering there was a positive association between growth, profitability and research ratios for the top 5 per cent. and bottom 5 per cent. of firms, but not throughout the whole range of firms. For the whole sample "additional research and development appears to make only a limited contribution to additional growth. The greater part of the differences between firms in rates of growth and profitability are due to other factors than differences in the amount of research and development done." [4]

The information in Diagram I is also rather limited. Correlations in this field are sensitive to definitions of industries. Industries as defined for statistical purposes vary greatly in their degrees of integration and it is possible for an industry with a high rate of growth in output and productivity to depend on the innovatory activities of its suppliers. Thus it matters greatly whether or not one defines the motor-car industry in Britain to include the firms which make the components.

However, the main problem here is one of interpretation. Should we conclude, or imply, that if the low growth industries had spent more on research and development they would have grown faster? The issue is not simply whether more physical output could have been produced but whether more output could have been profitably produced.

If economic growth opportunities were there but wasted by the industries with low percentage expenditures on research and development, this could mean that firms in such industries had failed to recognise (and/or exploit) opportunities for profitable investment in research and development. Or, it could mean that although the potential was there it was not possible for individual firms in the industry to realise it. This could happen where the firms in the industry were too small to finance and manage the research and development required, in which case state provision for research and development (as in agriculture), or cooperative research (as in steel, textiles, pottery, etc.), should solve the problem.

[4] *Industrial Research in Manufacturing Industry 1959–60* (London: Federation of British Industries, 1961), pp. 43–49.

There is little doubt that many firms in low-growth industries have often failed to recognise opportunities to invest in research. But whether British textile and metallurgical firms have been worse in this respect than paper firms (see Diagram I), I very much doubt. In any case one of the striking things about research and development percentages in different countries is that their industrial ranking is not very sensitive to the size of firms in the industries. This is implied in the following table which compares U.S. and British firms.

TABLE II

Research and Development as Percentage of Net Output, 1958 [5]

	U.S. Companies	U.K. Companies
Aircraft	30·9	35·1
Electronics	22·4	12·3
Other electrical	16·3	5·6
Vehicles	10·2	1·4
Instruments	9·9	6·0
Chemicals	6·9	4·5
Machinery	6·3	2·3
Rubber	2·7	2·1
Non-ferrous metals ...	2·0	2·3
Metal products	1·3	0·8
Stone, clay and glass ...	1·2	0·6
Paper	0·9	0·8
Ferrous metals	0·8	0·5
Food	0·5	0·3
Lumber and furniture ...	0·2	0·1
Textiles and apparel ...	0·2	0·3
All industries	5·7	3·1

The similarity in the pattern of industrial research and development between the U.S. and Britain is very striking. The only significant difference is in vehicles, which is partly explained by differences in industrial coverage. The main reason for the differences between industries is that the average profitability of research varies from industry to industry, depending on the state of technology and the extent of market saturation. It would clearly be quite inappropriate to conclude from the evidence of Diagram I that growth rates can be pushed up simply by raising research and development percentages.

[5] From Freeman, C., " Research and Development: A Comparison between British and American Industry ", *National Institute Economic Review* (May, 1962), 20, p. 31.

An appraisal of Ewell's " evidence " helps to throw further light on this issue. Ewell's forecasts proved to be very wide of the mark. His forecast was that in real terms research and development would rise by 27 per cent. and G.N.P. by 14 per cent. In fact, research and development rose by over 150 per cent. and became 2·8 per cent. of G.N.P., which rose by only 5 per cent.[6] In 1964 research and development will be more than 3 per cent. of G.N.P., although G.N.P. will be little above Ewell's prediction for 1960. Now it may be argued that this unexpected rise in research and development expenditures after 1954 could not yet have had its full effect on growth. This may prove to be so, but the fact is that growth has not risen to the levels expected from past research expenditure. In the period up to 1954, almost one half of America's cumulative expenditure on research and development had been in the last five years. Allowing a 5–10 year time lag, the growth effects Ewell confidently forecast from research should have shown at least from 1960 on.

The basic error in Ewell's approach was that he used a bi-variate approach to a multi-variate situation. He mentioned the importance of capital expenditure, production, sales, advertising, management, etc., but implied that we can simply take them for granted. He even calculated the yields to research and development as if they were the only costs of new innovation. Ewell did not discuss the objectives of research and development, whether certain types of research are more likely to have growth potential than others, the best mixtures of research and development, the alternative uses of scientists and engineers, or the possible tendencies to diminishing returns. By now it is clear that these are important factors, and that the model implied by Ewell—which is still frequently used—is far too simple to cope with the complex relations between science and growth.

We get a further indication of the complex relations between research and growth from the relations between G.N.P. per head and the percentage expenditure on research and development in different countries. These, which are shown in Diagram II, are sometimes taken to demonstrate the tendency of the R. and D. percentage to rise with G.N.P. per head.

A tendency can be weak or strong. The tendency shown in Diagram II is very weak. Japan is obviously well " off trend ". So too are Canada and Australia. Indeed, if a vertical line is run through the West German per capita G.N.P. it is plain that for six countries with similar levels of G.N.P. per head, R. and D. percentages range from 2·5 to 0·6. If, furthermore, at 1961 prices the U.S. per capita G.N.P. back to 1954 and the corresponding R. and D. percentages, were superimposed on Diagram II, we would see the

---

[6] The degree of error is so large that the revision of 1954 R. and D. expenditure, bringing it to 1·4 per cent. of G.N.P., does not make a great deal of difference.

R. and D. percentage move from 3 to 1 per cent. with scarcely visible change in per capita G.N.P.

DIAGRAM II [7]

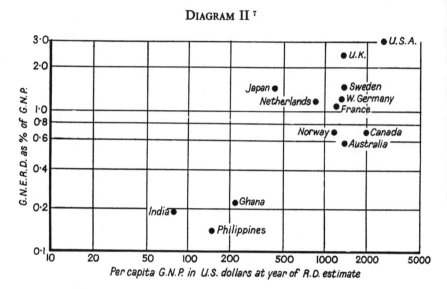

Per capita G.N.P. in U.S. dollars at year of R.D. estimate

After this discussion it should come as no surprise that internationally there is no sign of a high correlation between rates of growth in output per head and the percentage of G.N.P. devoted to research and development. The effect of plotting the research and development percentage against the annual percentage growth of national product per man for nine countries is shown in Diagram III.

It may be argued that plotting research and development rates against growth rates for the same years is rather meaningless. This is a valid criticism (which should also be applied to Diagram I) if relative research and development rates have changed significantly. Research and development statistics in most countries are not very accurate but such evidence as there is suggests that lagging scrambled research and development figures five or 10 years behind the growth figures would not materially alter the picture.

In any case, we have very little clear-cut evidence about the appropriate time lags, which are not the same for research and development and probably not the same in different countries and over time. It is true that some research will have a direct impact on technology. Thus, successful research into optimum firing conditions in pottery could be applied directly to operating conditions in tunnel ovens. But for basic and background research and most industrial research, the impact on technology, if any, is

[7] From *Science, Economic Growth and Government Policy* (Paris: OECD, 1963).

DIAGRAM III [8]

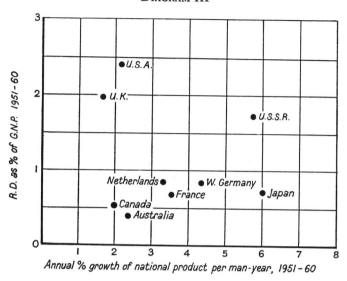

Annual % growth of national product per man-year, 1951-60

routed through the development process. This may be a very expensive business—as, for example, the building of prototype nuclear reactors of different kinds at Calder Hall, Dounreay and Winfreth Heath. It may also be a lengthy business. Although nuclear reactors have been built for the Central Electricity Generating Board to operate commercially they are not in fact competitive with traditional forms of power. It is probably wise to regard the nuclear stations as part of the development process required to get that experience of building and operation which may make nuclear power costs just break even with, and then undercut, conventional forms of generation.

Recent expenditure on development is more likely to affect contemporary growth than equally recent expenditure on research. In considering the impact of research and development on growth this might not matter if basic research, applied research and development occurred in fixed proportions. But they do not. The actual proportion appears to vary over time and between countries. Due to difficulties of classification, estimates of the research and development proportions in different countries are subject to wide margins of error and the figures given below should, therefore, be treated with some caution. Nevertheless, the possible errors of estimation are not large enough to destroy the general picture.

8 From Williams, B. R., *Investment and Technology in Growth* (Manchester: Statistical Society, 1964).

Table III

Estimated Percentage Distribution of R. and D. Expenditure in Four Countries [9]

	Basic	Applied	Development
United Kingdom (1961)	11	25	64
United States (1959)	8	22	70
France (1959)	26	74	
Australia (1961–62)	25	35	40

Research and development estimates are also scrambled in another important sense. The objectives of research and development also vary. It has been estimated that in the U.K. only 50 per cent. of research and development has a growth objective.[10] In so far as defence research and development (which until recently was over one-half the total expenditure in the U.S. and the U.K. and is now over one-third) contributes to growth it is in the nature of "fall out" or "spin off", which has not apparently been large. It is not then at all surprising that scrambled totals of expenditure on research and development have not shown a significant correlation with growth in national productivities.

Alternative Uses of Scientists and Technologists

It is now time to consider the implications of choice in the deployment of scientific manpower. Scientists and technologists are not only used in research and development. They are also used for the actual introduction and operation of new technologies. For example, in the chemical industry only about 50 per cent. of the scientists and 20 per cent. of the technologists employed are engaged in research and development. By contrast, in scientific instruments and electronic instruments the corresponding percentages are approximately 75 and 50. Given this sort of variation it is clear that the appropriate distribution of scientific manpower varies with industrial structure. But unless the distribution is appropriate and it would be very rash to assume that in each country it always is, growth potential will not be realised. From the need for scientists and engineers to introduce and operate sophisticated technologies there arises the possibility that research

9 From *Annual Report of the Advisory Council on Scientific Policy 1961–1962*, Cmnd. 1920 (London: H.M. Stationery Office, 1963); Keezer, D. M., " The Outlook for Expenditures on Research and Development during the Next Decade ", *American Economic Review*, L (May, 1963); *L'Usine nouvelle*, 18 May, 1961, p. 29; Williams, B. R., *Industrial Research and Economic Growth in Australia* (Adelaide: The Griffin Press, 1962).

10 Carter, C. F., and Williams, B. R., *Government Scientific Policy and the Growth of the British Economy* (Manchester School of Economic and Social Studies, 1964), reprinted from *Manchester School of Economic and Social Studies* (September, 1964).

and development may be excessive. It will be excessive, that is to say will actually hinder growth, when the output of potentially usable new technology is greater than the capacity of industry to absorb and when the use of more scientists and technologists outside research and development departments would increase industry's capacity to carry through technological change.

The possibility of excessive expenditure on research and development is increased by the fact that research and development do not have to be home grown. Unless a country is leading in all fields of science and technology, there must always be some choice between making at home and buying abroad. Now the less advanced a country's technology and the smaller its supply of scientific manpower, the greater the advantage of importing science and technology and of using a higher proportion of scientific manpower outside research and development activities.

Consider the Australian case. Because the Australian population is very small there is no chance of an effective research and development effort in more than a small part of the field. Furthermore, the stock of scientists and engineers per head of working population in Australia is only half that in Britain. It follows that the real cost of using scientists and engineers in industrial research and development and particularly in research is, in general, much higher in Australia than in England. In fact Australia achieves a higher growth rate with a very much lower research and development percentage than Britain and the United States. Part of the explanation is that the Australian industrial pattern is different. If, for example, each industry (including agriculture) devoted to research and development the British proportion of net output, Australian expenditure as a percentage of G.N.P. would be only (defence complications apart) 60 per cent. of the British level. But this is not the main part of the explanation. The science-based industries in Australia—vehicles, chemicals, electrical—are dominated by overseas companies and this, by reducing the need for home-grown research and development, increases the effective supply of scientific manpower. If in those industries the expenditure on research and development were at the British percentage of net output, research and development expenditure in Australian industry would be about 200 per cent. greater than it is. As it is, the subsidiary companies are able to economise research and development by adopting or adapting the technologies of the parent companies.[11]

Of course, there are other ways of acquiring new technologies from abroad. Licences to produce patented processes and products and the purchase of " knowhow ", play important roles in the diffusion of technology. Sometimes licence terms, even the grants of licences, depend on the ability

[11] See Williams, B. R., *Industrial Research and Economic Growth in Australia* (Adelaide: The Griffin Press, 1962).

to offer scientific or technological knowledge in exchange. But this is not the usual procedure. Thus, in 1961, the United States received from the recorded sale of licences and " know-how " $577 million, but paid out only $63 million. In 1963 Western Germany received DM200 million for patents, inventions and processes, but paid out DM550 million. Germany's payments for proven new technologies amounted to approximately one-sixth of its total expenditure on research and development, some considerable part of which will have no impact on growth. Clearly then the impact of foreign technology on German growth is even greater than the actual payment of DM550 million would at first sight suggest.

The following crude calculation helps to give some idea of the significance of this West German expenditure on foreign technological knowledge. British research and development is over $2\frac{1}{2}$ per cent. of G.N.P., the German just over 1 per cent. Now if we deduct the purely defence element in this and add in expenditure on foreign technological knowledge (multiplied by two in recognition of its proven impact on technology), the combined figure is approximately the same percentage of G.N.P. in both countries. If, as could well be argued is appropriate, a higher multiplier was used for expenditure on foreign " know-how ", the German percentage would be higher. But through these differing combinations of home-grown and imported technologies in the two countries the deployment of scientific and technological manpower differs greatly. The stock of scientists and engineers as a percentage of the labour force appears to be no lower in West Germany than in Britain.[12] It follows that Germany uses a considerably higher percentage of its qualified scientists and engineers outside research and development than does Britain. There is nothing in West Germany's record of postwar growth to suggest that this policy (which many scientists regard as parasitic) has not paid handsomely. Equally, there is very little in Britain's record of postwar growth to suggest that her policy has paid off.[13]

12 See *Resources of Scientific and Technical Personnel in the OECD Area* (Paris: OECD, 1963).
13 The failure of our massive expenditure on research and development to produce a dramatic change in the economic position has produced a certain gloom in " the scientific world ". But cheerfulness about the future keeps breaking in. For example: " . . . while spending on all R. and D. doubled between 1956 and 1962, spending on civil industrial R. and D. actually trebled. In this connection I would like to refer to a leader in *The Times* of 19 September, entitled ' Brighter News ', which discussed the recent encouraging rise in British exports and output. The writer sought an explanation of Britain's improving competitive position in the restraint of costs at home and in the increasing wage pressures on the Continent. While these may well be important and essential components, I would myself be interested to know how much our improved position can be attributed to this rapid increase in expenditure on industrial research and development to which I have just referred. Certainly we would expect that this investment would now be beginning to pay dividends." Sir Harry Melville, " Industrial Research and Development in Britain " in *Science and the City* (London: Harrison, Raison & Co., 1963), p. 41. Since Sir Harry wrote this, the performance of industrial production and exports has been very discouraging.

The Significance of R. and D. Percentages

There is no inherent virtue in high R. and D. percentages, just as there is no inherent virtue in high rates of capital expenditure. They are both parts of the investment process and as such constitute costs of growth. Optimum allocation of resources involves taking investment in R. and D., as well as in plant and equipment, to the point where prospective yields to additional expenditure equal the cost of the finance involved. The existence of average yields to research and development higher than the cost of additional finance does not as such indicate under-investment. The crucial thing is the marginal yield, which may quickly fall below the average yield because of the limited rate at which production and marketing departments can absorb new knowledge, instal new processes and market the additional (or different) outputs. This rate is itself affected by the distribution of scientific and technological manpower within the firm (and even on occasions within firms expected to buy the new products). Given the low elasticity of the supply of scientists, the time required to absorb newcomers, and possibly the need to recruit less able scientists and technologists, marginal yields to research in fields where both the market and technological potential for innovation is high will usually be very much lower than the average yields.

What can be said about the R. and D. percentage at which marginal yields will become equal to the cost of finance? Is there some basic tendency for the R. and D. percentage to grow with per capita G.N.P.?

We have seen that there are very wide differences in R. and D. percentages in countries with similar levels of per capita income and that national R. and D. percentages are not closely correlated with growth. Nevertheless, in almost every country with a growing standard of living, the ratio of R. and D. expenditure to G.N.P. has shown a strong upward tendency.[14] What is this due to and how far can we expect the rise to go?

Leaving aside military R. and D., the growth of which in the last five years has fallen behind civil R. and D., we can impute the rise in the R. and D. percentage to three factors. Because manufacturing industry generally has a much higher ratio of research to output than primary and tertiary industries, industrialisation, which brings an increase in the relative importance of the manufacturing sector, usually pushes up the R. and D. percentage. This, however, is a transitional influence and does not apply to countries which are highly industrialised. In such countries, a rise in the R. and D. percentage is due both to a shift within the manufacturing sector and to the " discovery " of research in the traditional industries.

A shift within the manufacturing sector (implied in Diagram I) towards the research-intensive industries is the effect of three factors. The first

[14] See *Science, Economic Growth and Government Policy* (Paris: OECD, 1963), pp. 22–23.

is the effect of industrialisation in poorer countries on the industrial structure of the richer countries. This industrialisation undermines the international trade position of the richer countries and forces them to specialise more in those industries which require higher rates of investment in plant and in people.[15] On the whole these are the research-intensive industries—though, as we have seen, part of the required research and development can be imported. The second factor is the tendency for people in rich countries not to react to their riches by demanding more leisure rather than more goods. This makes it profitable to invest in inventing new commodities. The third factor is the effect of higher incomes on investment in people, including the training of scientists and engineers. Here, as in the case of R. and D., the relation is very general. Thus, the Russian stock of scientists and engineers as a percentage of working population is higher than the British despite a considerably lower level of income per head. This sort of variation is indeed an important factor contributing to the confused position pictured in Diagram III. But given this greater supply of scientists and engineers, the capacity of countries to invent potential new products and processes and to use them efficiently is increased, so raising marginal yields to investment in technology.

There is, however, no reason to expect this process to continue unchecked. For, although some of the new processes and products created by research and development do not require capital expenditure to bring them into effect, many, perhaps most, of them do. Even if, as there is no reason to believe, the supply of very creative scientists and engineers increased proportionately to the overall supply, limits to the rate of growth in capital expenditure would set limits to the growth of research and development. In the United States, Professor Yale Brozen has argued that the correlation between research expenditures and subsequent sales has tended to weaken since 1957 and this is a sign of research and development expenditure moving nearer to an equilibrium situation. In 1960 he predicted that the portion of national product devoted to research and development would rise from 1·9 to 3·0 by 1975 with a ceiling of 5 per cent. in the next century.[16] His 3 per cent. has arrived a decade early, partly because, like Ewell, he overestimated the growth of national product. This is an important point, for nothing pushes up research and development percentages faster than a failure to achieve the growth expected from research.

15 See *Industrialization and Foreign Trade* (Geneva: League of Nations, 1945).
16 Brozen, Yale, " The Future of Industrial Research and Development " in *The Rate and Direction of Inventive Activity*, pp. 273–276.

What Should We Expect from Research

Countries differ in levels of income, in levels of technology, in industrial structures, in size, in the proportion of scientists and engineers in the work force and so on. These differences affect the R. and D. percentages at which the yields on additional research fall to a low level; the ratios of home-grown to imported science and technology; the appropriate distribution of expenditure between basic and applied research and development and of manpower between research, development and other activities.

In a country such as the United States, with a relatively high proportion of scientists and engineers in the work force and a technological lead in most fields, a high R. and D. percentage will be consistent with an attempt to achieve high growth and a condition of it. A high R. and D. percentage will not leave a critical shortage of men required to instal and manage new technologies. And given the restricted opportunity to borrow new technologies from abroad, a big development effort will be required to create the new technologies. As the country gets richer there is likely to be an increasing struggle to maintain a high growth rate as trade positions in existing export products are weakened by the diffusion of its new technologies abroad and higher investment rates are required to establish the still newer technologies. Just how all this will affect the total relations between R. and D. percentages and growth will depend on a very large number of factors, including trade policy, population growth, military expenditures and future income elasticities of demand for leisure. If, for example, the demand for leisure increased, we could expect relatively greater emphasis on labour-saving process innovations, which would have a bigger effect on product per man hour than on product. In other words, product per man hour could grow rapidly, even though G.N.P. did not.

Between the countries of Western Europe and the United States there are important differences, which should affect what we expect from research. The fields where any country has a technological lead over the U.S. are few and in all countries the stock of qualified manpower is a smaller percentage of the work force than it is in the U.S., and very much smaller anyhow. It follows that no country can afford to try to cover the whole field of industrial research and development [17]; that any such attempt must lead to a weak effort and, therefore, a low R. and D. yield in all fields. It follows too from the possibility of importing technology from the U.S. —whether in the form of payment for technical knowledge or the willingness to allow the operation of American subsidiaries—that the dependence of

---

[17] Indeed, in some cases even one field is too expensive. Hence the collaboration between France and Britain in developing the Concord and between the Common Market countries in the development of nuclear power.

innovation on home research and development is less than in the U.S.[18] Just how far it will pay a West European country to import technology will depend on the cost of acquiring it and the efficiency of home R. and D.[19] It is thought, for example, that the recent rapid rise in Japanese research and development has been connected with a rather steep rise in the cost of buying " know-how ".

What follows about the distribution of qualified manpower between research and development and other activities is not so clear. For, as we saw in the case of Australia, the use of foreign subsidiaries in research-intensive industries has the effect of adding to the effective manpower supply. But the probability is that the most economic use requires a smaller proportion of scientists and engineers in research and development than in the United States.[20] On the other hand, purchasing " know-how " and giving foreign firms with advanced technologies facilities to operate, may have a greater economising effect on development than on applied research. The precise effect will, of course, depend on the fields in which most use is made of foreign development work. But certainly the effect could be to produce more research in relation to development than would be sensible in a closed economy.

[18] This is one reason why correlations between R. and D. and profits are higher in the U.S. than in Britain.
[19] For an account of the great impact of U.S. firms on British innovation see Dunning, J. H., *American Investment and British Manufacturing Industry* (London: George Allen and Unwin, 1958).
[20] Some of the implications of all this for government scientific policy in Britain have been examined in Carter, C. F., and Williams, B. R., *op. cit.*

SCIENTIFIC CHOICE AND BIOMEDICAL SCIENCE

ALVIN M. WEINBERG

I CONTEND in this paper that of all the sciences now supported by our society, biomedical science ought to stand first. We are, or ought to be, entering an age of biomedical science and biomedical technology that could rival in magnitude and richness the present age of physical science and physical technology. Whether we shall indeed enter this age will depend upon the attitude toward Big Biology adopted by biomedical scientists and governmental agencies that support biology. Whether the age of Big Biology will be truly rewarding will depend on the common sense and integrity of all who participate in this adventure.

The Ongoing Debate

The scientific-political world has been debating scientific priorities with growing zeal during the past three or four years; yet in this debate the voice of the biologist has been rather mute. The public debate began, informally, with a number of essays on scientific choice by American and English authors.[1] Since then the debate has become more formal and quite widespread. For example, in the United States, the Committee on Science and Public Policy (C.O.S.P.U.P.) of the National Academy of Sciences has sponsored reports by groups representing different branches of science; these reports summarise the achievements, promise and needs of particular branches of science. Such " planning reports " on ground-based astronomy [2] and chemistry [3] have already appeared. Similar reports on physics, computers, mathematics and botany are being prepared. In the biomedical sciences a comparable effort under the leadership of Professor Philip

[1] Carter, C. F., " The Distribution of Scientific Effort ", *Minerva*, I, 2 (Winter, 1963), pp. 172–181; *idem*, Letter to the Editor, *Minerva*, II, 3 (Spring, 1964), pp. 382–383; Dedijer, Stevan, Letter to the Editor, *Minerva*, III, 1 (Autumn, 1964), pp. 126–129; Maddox, John, " Choice and the Scientific Community ", *Minerva*, II, 2 (Winter, 1964), pp. 141–159; Toulmin, Stephen, " The Complexity of Scientific Choice: A Stocktaking ", *Minerva*, II, 3 (Spring, 1964), pp. 343–359; Weinberg, Alvin M., " Criteria for Scientific Choice ", *Minerva*, I, 2 (Winter, 1963), pp. 159–171; *idem*, Letter to the Editor, *Minerva*, II, 3 (Spring, 1964), pp. 383–385; *idem*, " Criteria for Scientific Choice II: The Two Cultures ", *Minerva*, III, 1 (Autumn, 1964), pp. 3–14.

[2] *Ground-based Astronomy: A Ten Year Program*. A Report prepared by the Panel on Astronomical Facilities for the Committee on Science and Public Policy of the National Academy of Sciences (Washington: National Academy of Sciences—National Research Council, 1964).

[3] *Chemistry: Opportunities and Needs*. A Report on Basic Research in U.S. Chemistry by the Committee for the Survey of Chemistry, National Academy of Sciences—National Research Council (Washington: National Academy of Sciences—National Research Council, 1965).

This article from *Minerva*, IV, 1 (Autumn, 1965), pp. 3–14.

Handler is just getting under way. Other reports such as those on earth sciences [4] and high energy physics [5] have also been published.

Despite the value of these formal reports, I think it is important that the informal debate on scientific priorities continue. Formal reports delineating the achievement and promise of various fields all tend to be isomorphic. It makes little difference whether the field is astronomy, physics, or computers: its achievements have been outstanding, its promise superb and its needs and tastes very expensive. Nor is this surprising. Each report is prepared by dedicated members of a particular scientific community whose passions and aspirations, as well as knowledge, centre on a single field. The very reasonable theory underlying the preparation of these reports is that each field should put its very best foot forward. Judgements among the fields would then be made by a higher body, like the President's Science Advisory Committee, that represents many different scientific fields.

Actually, the political process out of which flows our ordering of priorities does not work that neatly. Though the Science Adviser carries great weight, Congress and the separate government agencies must also be reckoned with and their views are harder to bring into focus. Interpretative and philosophic analyses of the problem of scientific choice, particularly judgements as to relative priority, will therefore remain important. Such judgements, by the nature of things, can hardly be other than individual opinions. Out of such individual views and opinions is fashioned a climate of thinking, an intellectual environment, which impinges in countless small ways on those in Congress and in the agencies who make scientific policy.

Some such view of the nature of the problem of choice, as viewed by Congress, was implicit in the response by the Committee on Science and Public Policy to two questions asked recently by the Subcommittee on Research and Development of the House Committee on Science and Astronautics. In effect, the Chairman of the Subcommittee, Congressman Daddario asked first, how much science our society should support; and second, how should the total science pie be cut? Rather than hammering out a weak consensus to such loaded questions, Professor George Kistiakowsky, former Chairman of C.O.S.P.U.P., asked each member of the committee to prepare an essay for which he alone was responsible, although each essay was criticised by other members of the group. This

---

[4] *Solid-Earth Geophysics:Survey and Outlook.* Panel on Solid-Earth Problems of the Geophysics Research Board and Division of Earth Sciences, National Academy of Sciences—National Research Council (Washington: National Academy of Sciences—National Research Council, 1964).

[5] *Report of the Panel on High Energy Accelerator Physics of the General Advisory Committee to the Atomic Energy Commission and the President's Science Advisory Committee,* TID-18636 (Washington: U.S. Atomic Energy Commission, Division of Technical Information, 1963).

way of dealing with a question of public policy preserves the congressional tradition of eliciting many different opinions in arriving at a course of action. The collection of 15 essays on scientific choice is, I believe, a useful contribution to the debate on allocation of resources to science.[6]

The Argument for Biomedical Research

Any judgement as to the relative worth of any field of human activity involves an assessment of how that activity bears on human values. In particular, we support large-scale science because, in one way or another, we believe that out of large-scale science will come human benefits or values. Now the *value* of science cannot be determined from within science. It is a venerable philosophic principle that the value of any universe of discourse must be judged from outside that universe of discourse. It was for this reason that, in an earlier article on scientific choice,[7] I urged that large-scale public support be given a field of science *only if* it rated well with respect to what I called " external criteria ". These I identified as technological merit (meaning bearing on related technology), scientific merit (meaning bearing on related fields of science) and social merit (including national prestige, culture, etc.).

Of all the bases for claiming large-scale public support of a scientific activity, the possibility of alleviating human disease through such activity is one of the most compelling. Of all the sciences, the biomedical sciences are most directly aimed at and most relevant to alleviating man's most elementary sufferings—disease and premature death. There is urgency of the most excruciating kind in getting on with this job. The assault on human disease, insofar as it may result in alleviation of immediate everyday human suffering, has an urgency about it comparable to the urgency with which a nation prosecutes a war. Indeed, I would draw an analogy in this regard between war-time research in physics and present-day research in the biomedical sciences.

This claim to urgency can hardly be matched by any of the other great fields of natural science. Certainly those fields that base their claims to support primarily on the promise of enlarging the human spirit have, to my mind, a less valid case for *urgency* than do those fields that base their claim on the possibility of curing or preventing human disease. SU(n) symmetry is magnificent and soul-satisfying to those who understand it; a cure for leukaemia is more immediate in its benefit to mankind.

[6] *Basic Research and National Goals. A Report to the Committee on Science and Astronautics, U.S. House of Representatives, by the National Academy of Sciences* (Washington: Government Printing Office, 1965), ix + 336 pp.
  Two of the papers by members of the Committee on Science and Public Policy— " Federal Support of Basic Research: Some Economic Issues " by H. G. Johnson and " Scientific Choice, Basic Science and Applied Missions " by A. M. Weinberg—were reprinted in *Minerva*, III, 4 (Summer, 1965), pp. 500–523.
[7] Weinberg, Alvin M., " Criteria for Scientific Choice " *Minerva*, I, 2 (Winter, 1963), pp. 159–171.

Are the biomedical sciences that relevant to the conquest of disease? To an applied scientist like me, this question seems absurd. What strikes an observer most about modern biology is how the new viewpoints have unified the subject. The genetic code appears to be universal. The dogma of protein synthesis—DNA, messenger RNA, transfer RNA, protein—seems to be valid in almost every life form. The same 20-odd amino acids build proteins in bacteria, in mice, and in men. This unity suggests that most of what we learn about biological mechanisms in almost any animal is likely to have ultimate medical applications, whereas the same degree of relevance to application cannot be claimed for large parts of modern physics, or astronomy, or mathematics. In the biomedical sciences the distinction between pure and applied is rather irrelevant. The distinction is better made between intelligent, imaginative research and unintelligent, plodding research. As a matter of tactics, I have therefore argued that all of the biomedical sciences be viewed as applied science, even though I know that calling some of my good friends who consider themselves to be basic biological scientists " applied " scientists hardly endears me to them. Yet from the point of view I am discussing here—the validity of the biomedical sciences' claim to urgency and therefore the validity of their claim to large-scale support from society—the position of biology is far stronger if it regards itself as fighting the war against disease instead of the war to enlarge the human spirit, worthy as the latter is.

If the biomedical sciences are viewed as applied sciences, aimed at alleviating disease, then in assessing their priority they should be judged not so much against other branches of science that are not aimed at the same goal but rather against alternative means of alleviating disease. The most obvious such alternative is medical practice, including treatment centres, hospitals, medical education, nursing care, etc. And indeed, I believe there is evidence of competition between the demands of medical practice and the demands of medical research. I refer to the frequently quoted statistics showing that the relative number of A students in first-year medical school in the United States fell from 40 per cent. to 13·4 per cent. during the period from 1950 to 1960.[8] Although it is hard to document, I have always believed that at least part of this loss in quality was a consequence of the favoured position of the graduate student in biomedical research as compared with his counterpart in medicine. The United States Government has made fellowships available for the research student but, with few exceptions, not for the medical student. I expect this situation to change in the United States as a result of such studies as that by the President's Commission on Heart Disease, Cancer and Stroke, which bring

8 Wiggins, Walter S., et al., " Medical Education in the United States ", Journal of the American Medical Association, CLXXVIII, 6, 11 November, 1961, p. 601.

the country's attention to the need for more medical practice. My own view is that we need more biomedical science *and* more medical practice and that the two, taken together, deserve very high priority in the allocation of resources.

The Prospect of Returns from Biomedical Research

Relevant as is the aim of a science to achievement of a recognised human value—in the case of biology to the elimination of human disease —this can only be a partial justification for large-scale public support. Before any scientific field can expect support on a very large scale it must be at a stage where large-scale public support is likely to produce useful results. Anyone who claims that biomedical science should become our number one scientific priority must show that this field is likely to give fair return for support received.

In this respect the situation in the biomedical sciences at first sight seems to stand between certain of the physical sciences and the behavioural sciences. Judging by the criterion of direct relevance to human welfare, any ordering would almost surely place the behavioural sciences at least on a par with, if not above, the biomedical sciences; the more abstract physical sciences would almost surely rate below these. Judging by the criterion of intellectual readiness for exploitation—*i.e.*, whether it is a lack of large-scale support which is mainly responsible for restraints on progress—abstract physical science, like elementary particle physics or astronomy, is probably ahead of biomedical sciences and the behavioural sciences are much farther behind. This at least is the view one would gather from the strength of the plea for support made by the physical scientists, compared with the relative weakness of the plea we hear from the biomedical scientists. I think, however, that the biomedical scientists understate their case.

To begin with, the war on human disease is a tangible war—more tangible, say, than our efforts to enlarge the human spirit—and it should be fought with the same attitudes we adopt when fighting a real war. We expect smaller returns per dollar expended when fighting a war than when carrying on a less crucially important activity. So I would argue that, because of the importance of each victory in the battle against disease, we ought to be willing to get less per dollar spent on biomedical research than we are willing to get from expenditures on the more remote fields of science. We should stop putting more resources into the enterprise only when we have reached a stage of negative returns—when more resources *reduce* the total useful output—not merely raise the unit cost of an increased total output.

I believe the biomedical sciences are not near the stage where additional large-scale support will *reduce* the total output of the entire enterprise. It is apparent even to the most casual observer that we are beginning to understand many of the life processes which have been mysteries for so long: the revolution in molecular biology, including the unravelling of genetic codes, determination of the structure of proteins and insights into enzyme action; or the beautiful elucidation of the mechanism of nerve action; or the new insights into the genetic control of immune mechanisms; or the extraordinary implication of viruses in some cancers, notably animal leukaemia, although their role had long been suspected. One can hardly believe that the many fruitful points of departure uncovered during the past decade are anywhere close to being exploited; or that, if more well-trained, well-supported, investigators were set to work, new and startling points of departure would not emerge.

Moreover, the biomedical sciences can be force-fed, even more than they are now being force-fed. More money for biology has raised the salaries of biologists, at least in the United States, so that now the biologists enjoy an unaccustomed affluence. Though this state of affairs annoys administrators, particularly of multi-disciplinary laboratories where disciplines use each other's salary schedules as ratchets, the overall effect as far as biomedical science is concerned is, on balance, good. More intelligent young men and women are attracted to well-paid careers than to poorly paid ones. In the United States such force-feeding of a discipline in the past has produced results. For example, the Atomic Energy Commission, by pouring money into nuclear research, caused nuclear research to flourish and encouraged many young science students to go into nuclear research. Or again, the Atomic Energy Commission and the U.S. Department of Defense deliberately established about a dozen interdisciplinary materials research laboratories; though it is too early to say positively, my impression is that materials research in the United States has profited by this action.

The Absorption Capacity of Biomedical Research

There are other reasons, intrinsic to the changing style of research in biology, why more money will be needed. Most obvious is the growing cost of equipment. A modern electron microscope now costs $40,000 and more and more cellular biology seems to depend on the electron microscope. Even now attempts are being made both at Oak Ridge National Laboratory and at Argonne National Laboratory to develop an electron microscope with a resolution of $1A°$. Such a device, if successful, would enable one to identify individual atoms in biological molecules. It could cost several million dollars.

But there are other, possibly subtler, reasons why biological research is becoming more expensive and is requiring more people. In earlier times, when biology was Little Science *par excellence*, biologists were content to look only at those problems that could be handled by the style of Little Science. Genetics was done with fruit flies, with their large chromosomes, because fruit flies are inexpensive, not because fruit flies are as much like man as are mammals. Those questions that required large protocols of expensive animals were answered poorly or not at all—not because the questions were unimportant but because to answer them was expensive and required the style of Big Science which was so foreign to the biologists' tradition.

But this is changing, in part at least, because the Big Scientists from neighbouring fields have taught the sin of Big Science to the biologists. Perhaps the best known example of the drastically changed style of some biological research is the large-scale mouse genetics experiment of Dr. W. L. Russell at Oak Ridge. For the past 16 years Russell has been studying the genetic effects of ionising radiation in a mammal, the mouse. Since mutations even at high dose rates are so rare, Russell uses colonies containing 100,000 mice. To perform such experiments takes much money and many people; and yet it seems impossible to visualise any other way of obtaining the data.

The problem of large protocols which Russell faced and the Atomic Energy Commission solved (at a cost of 10^6/year for this single experiment) is one which arises in many other situations. The increasingly important matter of low-level physical and chemical insults to the biosphere will require many large experiments if we are to assess accurately the various hazards that now bombard us. Or take old age, the commonest " disease " of all: merely because the effects are subtle and often appear haphazardly, the study of aging requires large and expensive protocols. The tradition of the biologists, and it is a very honourable and desirable tradition, has been a niggardly one; biomedical research avoided expensive experiments even if expensive experiments were required to obtain reliable statistics. I believe that biology, while continuing its tradition of Little Science, shall have to accept also the style of Big Science and that, even though this is expensive, the biologists will find the public willing to support them.

There is another trend in the style of biology which will add to its expense. This is the increasingly interdisciplinary character of modern biology and, particularly, its increasing dependence on the techniques and

113

methods of the physical sciences and even of the engineering sciences. A few examples, taken from our own experience at Oak Ridge, will illustrate these points. For example, in attacking the problem of radiation insult, we have mobilised biochemists, cytologists, geneticists, pathologists and biophysicists. Our dependence on disciplines even farther removed from biology is growing. Thus, our biochemists, notably Dr. G. D. Novelli and Dr. M. P. Stulberg, need large quantities of t-RNA, preferably separated into unique fractions, to study how amino acids are assembled into proteins. The problem in many ways is one in chemical engineering and some of the chemical engineers, particularly Mr. A. D. Kelmers, at Oak Ridge National Laboratory have pitched in to help. What the chemical engineers have done already strikes me as being rather impressive. They have been able to extract as much as 600 grams of pure t-RNA from 300 kilograms of E. coli by fractionating crude nucleic acids in a sodium acetate-isopropanol mixture followed by selective elution from a DEAE-cellulose column. They have then fractionated the specific t-RNAs by using a liquid ion exchange system based on quarternary ammonium compounds of the general sort developed at Oak Ridge National Laboratory in refining uranium ores. The resulting separations are superior to any that have been achieved by older methods.

Second, I mention, again from Oak Ridge experience, the exciting developments in zonal centrifugation applied to biology. For many years very high speed, very large, continuously fed centrifuges have been developed for separating the isotopes of uranium. Much of this work has been carried out at the K-25 Gaseous Diffusion Plant. Some four years ago, Dr. N. G. Anderson of the Oak Ridge National Laboratory Biology Division realised that such centrifuges, suitably modified, might separate cellular moieties on a larger scale than could be done with any other technique. And indeed, with the generous support of the National Cancer Institute and the Atomic Energy Commission, this is exactly what has happened. With these centrifuges Anderson has been able to detect viruslike particles in leukaemic blood more consistently than have most other investigators who do not have this tool available. I would expect Anderson's centrifuges to become widely used in biomedical research, even though some of his centrifuges cost as much as $45,000.

I could list many other instances of the growing interaction between the biological sciences and the physical and engineering sciences—for example, the technique of medical scintillation spectrometry which has become a medical specialty in its own right; or the wide use of computers in biomedical science; or, for that matter, the application of the methods of quantum chemistry to the attempts to understand the carcinogenic action

114

of aromatic hydrocarbons. But I have given enough examples to bring out the main points: that biomedical science is becoming even more inter-disciplinary; that the disciplines and techniques it draws upon are expensive; and that this will add to the expense of biomedical science.

The Division of Labour between Universities and Research Institutes

The changing style of biomedical research and its great and urgent expansion will affect the future organisation of such research. At present, a very large part of biomedical research is carried out at universities— institutions that are, or should be, committed to education at least as strongly as they are committed to research. University biomedical research must flourish and, to do this, it must grow. We shall have to maintain Little Biology as well as Big Biology and we shall have to produce many more trained biomedical scientists if we are to attack, with either style, the problem of human disease with sufficient urgency.

But much of the great expansion in biomedical research should take place in biomedical research institutes, many of which will be directly affiliated with universities, but many of which will not. For, as Professor Rossi put it so well in a recent issue of *Daedalus*,[9] the social ecology of the university is not as well suited to a massive attack aimed at a single goal as is the ecology of the research institute. In the first place, the traditional departmental structure of the university is poorly suited to interdisciplinary approaches. In the second place, in the university indi-viduality and academic freedom are preciously guarded prerogatives and these are often incompatible with achieving success in tasks that require cooperation.

The ecology of the research institute has a different tone: it is more hierarchical, its members interact with one another more strongly, and it is interdisciplinary. In the individualistic, competitive university environ-ment, genius flourishes but things go slowly because each genius works by himself with his own small group of students and assistants. In the less individualistic, cooperative institute environment, genius probably does not flourish as well but things go very fast because so many different talents can be brought to bear on a given problem. It is a place in which, however, a single, very able man can exert much more power and influence than he can in the university environment; it is a place where the whole is often much more than the sum of its parts.

If one accepts the proposition that biomedical science ought to be pursued with the same urgency with which we pursue military research,

---

[9] Rossi, Peter H., " Researchers, Scholars and Policy Makers: The Politics of Large Scale Research ", *Daedalus*, LXLII, 4 (Fall, 1964), pp. 1142–1161.

then the institute provides a better setting for such activity than does the university. In speaking this way I admit to being very much influenced by our own experience at Oak Ridge. There we have a prototype of a large biomedical institute: its central theme is the radiation insult to the biosphere. In pursuing this major theme, many disciplines are brought to bear. The enterprise is benevolently hierarchical; it is large; it is interdisciplinary; and I think it is effective.

I would therefore suggest that much of the big expansion in biomedical research ought to go toward establishing additional interdisciplinary institutes, like the Sloan-Kettering Institute, or the contemplated environmental health institute of the World Health Organisation. Certainly close connections with the universities are desirable; but I do not regard these as primary. The main job is to learn as much as possible in as short a time as possible to alleviate human suffering. In some cases this aim is furthered by close association with a university. I suspect that there are many cases where only a loose university affiliation is desirable.

Collaboration with the Physical Sciences: Financial Aspects

The coming age of biomedical science will impose on administrators of biomedical research a new and unaccustomed responsibility toward the physical sciences. I have already alluded to the increasing relevance of the physical sciences to the biomedical sciences. It is time for the community of biomedical science to recognise its dependence upon certain of the physical sciences and to assume a proper share of their support.

Support of certain parts of physical science has already been taken up by the biomedical sciences. For example, in the United States, the National Institutes of Health are now the largest single supporters of basic chemical research in the universities. But my impression is that such support tends to be somewhat constrained by narrow interpretations of relevance.

Research in many of the physical sciences—like structural organic chemistry, or X-ray and neutron diffraction, or even certain parts of solid state physics—is the proper concern of the biological sciences. The whole Watson-Crick development would have been impossible had it not been for major developments in the techniques of X-ray diffraction. Moreover, more and more of the world's leading biologists seem to be coming from the physical sciences: I mention, for example, Dr. Francis Crick, or Professor Seymour Benzer, or Professor Paul Doty, or Dr. Kenneth Cole. The debt owed to the physical sciences by the biomedical sciences is one of long standing and it is growing. It is now time for the biomedical sciences to begin repaying this debt.

The basic physical sciences in the United States are facing a major financial crisis. In the past they have been supported largely by three agencies: the Department of Defense, Atomic Energy Commission, and National Aeronautics and Space Administration. But the missions of these three agencies—defense, atomic energy and exploration of space— are not likely to receive increasing support; on the contrary, the United States in the past year has made the political decision to keep these agencies at about their present level, or even to reduce them somewhat. Thus the physical sciences, insofar as they are supported because they are relevant to the achievement of the missions of these agencies, are probably destined to receive relatively less support in the future than they have in the past.

But this predicament comes at the time when support for the biosciences should greatly increase and when the connections between the physical and the biomedical sciences become ever stronger. What is more natural than to ask the biomedical sciences to carry a fair share of the burden for supporting the many branches of physical science that are broadly relevant to the biomedical sciences? Such a plea from the hard-pressed physical scientist has justice on its side. The biomedical administrators, in their newly found affluence, should heed these cries from their colleagues in the physical sciences who have helped them so much for so many years.

Big Science and Little Science in Biomedical Research

Traditional biologists must surely recoil in horror at the advice given here:—to expand even at the cost of individual effectiveness as long as their total output increases; to break down their traditional disciplinary barriers and to adopt more of the institute, as contrasted to the university, style of research; to overcome their suspicion of the physical scientists; in short, to accept the new style of Big Science, in addition to the old style of Little Science.

If this is their reaction, they should be reminded that insofar as what they do is part of the war against human suffering their desires and tastes are not all that matter. Biomedical science is not done or, more importantly, not supported by the public simply because it gives intense satisfaction to the dedicated and successful biomedical research worker. It is supported on a really large scale because out of it have come means of eliminating man's infirmities. If a style that complements the traditional style is needed in order to build a much larger biomedical research enterprise, then this style will have to be adopted much as it hurts the sensibilities of those attached to traditional patterns of scientific organisation.

117

I have myself inveighed against the dangers of Big Science:—its preoccupation with the grandiose announcement rather than the great discovery; its substitution of money for thought; its over-abundance of administrators; its incompatibility with the educational process; even its inefficiency. As Sir Winston Churchill once said, "I do not unsay one word of this". But nothing I have said implies that I consider the style of Little Science to be obsolete. In urging more biomedical science, I plead both for more Big Science and for more Little Science.

Big Science, with all its dangers, does have a real place in the scheme of things. When the end to be achieved is important enough, and when the state of the science suggests that more support will lead to more results (and both these circumstances apply to biomedical sciences), then we are justified in going all out in our plea for public support. More than that, we have a responsibility to apprise the political leadership of the country of this belief. The coming age of biomedical science will not be an unmixed blessing for the biologist: he surely will fret at being involved in something big and unwieldy and at times inefficient. Nevertheless, as a responsible member of the human race who is sensitive to the purpose of enlightened human activities such as biomedical research, he will have to submerge his instinctive distaste for bigness in the interest of the welfare of humanity.

THE COMPLEXITY OF SCIENTIFIC CHOICE II:
CULTURE, OVERHEADS OR TERTIARY INDUSTRY?

Stephen Toulmin

The problems debated nowadays under the heading of " science and public policy " are in some respects old, in others entirely new. In an academic way, the questions they raise have been debated ever since the seventeenth century—if not before. The foundation of the Royal Society of London under the wing of King Charles II's Admiralty anticipated by nearly 300 years the patronage which pure science has won in our own generation from the United States Office of Naval Research; and Bishop Sprat[1] could write as persuasively as Jerome Wiesner[2]—and in even more sonorous language—about the technological " spin-off " to be expected from this royal patronage. So, at the very beginning of the modern scientific period, two rival motives were injected into the debate about the role of science within the larger society or commonwealth. The prototypical figures, standing for these two motives, are Francis Bacon and Isaac Newton—Bacon, who emphasised the value of science in making two blades of grass grow where only one had grown before, and Newton, in whose life and thought alike the philosophical values of science far outweighed the technological ones.

In the last 10 years, however, this particular debate has lost its purely theoretical air and acquired a new practical urgency. Only a few years back, recognition of the " exponential growth "[3] of science was leading its spokesmen to emphasise the great contribution which the increasing range of scientific work could make to the community at large. By now, scientists themselves are beginning to talk less optimistically. The growth-curve appears to be not exponential but S-shaped. As a result, the questions arise: have we the manpower? Are we past the inflection-point? Will national governments in general (and the United States Federal Government in particular) consent to go on paying vastly increasing sums to support the research of men who promise no immediate product or invention? In short: will society continue to subsidise the work of men whose prime

[1] Sprat, Thomas, *The History of the Royal Society of London, for the Improving of Natural Knowledge* (London: Printed by T.R. for J. Martyn and J. Allestry, 1667).

[2] Wiesner, Jerome B., *Where Science and Politics Meet* (New York: McGraw Hill Book Co., 1965).

[3] Price, Derek J. de Solla, *Science since Babylon* (New Haven: Yale University Press, 1961) and *idem, Little Science, Big Science* (New York: Columbia University Press, 1963).

This article from *Minerva*, IV, 2 (Winter, 1966), pp. 155–169.

interests are intellectual, not technological, and whose final loyalty is in many cases to some ideal supranational community of scholars, rather than to any individual corporation or country? Hitherto, the growth in the scale of activity in pure science has not been uneventful but it has been unbroken. Now, for the first time, the questions have become very real ones; where are the men and the money to go on coming from, if scientific growth is to continue? And what kind of case must scientists put to their fellow citizens—in particular, to their political representatives—if they are to have continued support from government for their pure research?

Against this background, it is not surprising that the old duality in the arguments for science—philosophical and technological—should come to the surface and be seen as posing difficult questions. The resulting dilemma has recently been argued in print by Dr. Alvin Weinberg[4] who puts the difficulty in the form of the question: " Should governments support basic science as a branch of high culture, or as an overhead charge on applied science and technology?" He ends by concluding that—noble aspirations apart—the " overheads " argument must at the present time prevail; and this conclusion also forms the basis of the report on *Basic Research and National Goals*, presented by a committee of the National Academy of Sciences (N.A.S.) to the Committee on Science and Astronautics of the U.S. House of Representatives.[5] The contrary view, that—over and above any question about overheads—national governments must be prepared to support science as an essential part of a high civilisation, was argued by a panel under the chairmanship of Dr. Glenn Seaborg in its report to the President of the United States.[6] The purpose of the present paper is to call this basic dichotomy in question. In our actual situation, and even more in the situation into which we are moving—it will be argued—this opposition between " science as culture " and " science as overheads " is no longer a helpful, relevant, or even a *legitimate* one. We are, rather, compelled by the facts of late twentieth-century life to move beyond it.

The paper is, accordingly, in two parts. First, we must look carefully at the respective strengths and weaknesses of the " overheads " and " high civilisation " doctrines, as they appear to us today. Then, in the second half of the paper, I shall argue that we are now at a transition point in economic and social history, the effect of which will be to make this

---

[4] Weinberg, Alvin M., " Criteria for Scientific Choice II: The Two Cultures ", *Minerva*, III, 1 (Autumn, 1964), pp. 3–14.

[5] *Basic Research and National Goals: A Report to the Committee on Science and Astronautics, U.S. House of Representatives by the National Academy of Sciences* (Washington: U.S. Government Printing Office, 1965).

[6] *Scientific Progress, the University and the Federal Government.* Statement by the President's Science Advisory Committee, prepared by a panel on " Basic Research and Graduate Education " under the chairmanship of Dr. Glenn Seaborg (Washington: The White House, 1960).

apparent opposition progressively more unreal. Before the end of the twentieth century (I shall try to show) new patterns of society and employment which are already emerging today will have turned scientific research—even of the most basic, "cultural" kinds—into a "tertiary industry", the social and economic values of which go far beyond all questions of "productive output", "technological spin-off" and the like.

The Overheads Doctrine

In the long run (says Dr. Weinberg) we can dream of a world in which the *cultural* significance of science would be fully appreciated: in which every educated citizen and every influential politician would have *some* first-hand feeling for pure science, as well as for the fine arts—responding to the beauty of (say) Schrödinger's equations or the DNA helix as he might to Bach, J. P. Souza or the Beatles. In such a world we would all, after our own fashions, be either scientists or patrons of science; and then the political case for supporting pure science would not be so urgent. But this world (he concludes) is, after all, only a dream:

> It is probably Utopian—as much as Shaw's *Back to Methuselah*—to expect every man in the street to become an amateur scientist or even a science fan.[7]

In the real world, and for the foreseeable future, the *political* case for government support of pure science cannot afford to be based on a cultural view of scientific research:

> Until and unless our society acquires the sophistication needed to appreciate basic science adequately, we can hardly expect to find in the admittedly lofty view of 'science as culture' a basis for support at the level which we scientists believe to be proper and in the best interests both of society and of the scientists.[8]

Rather, we must view the role of pure science on the national scene in the same general spirit that the directors of the Bell Telephone System view the basic science done in their own research laboratories. No specific financial investment made into basic research can be absolutely *relied* on to pay technological dividends; but some proportion of the research budget must be invested in basic research. Nothing venture, nothing win. If you don't speculate, you'll never accumulate.

On a more extensive scale, the same argument can be found in the National Academy of Sciences' report on *Basic Research and National Goals* presented to the United States Congress in March, 1965. The specific function of this report was to discuss the principles which should govern

---

7 Weinberg, Alvin M., *op. cit.*, p. 12.
8 *Ibid.*, p. 14.

federal support for basic research and the greater part of the report took the form of 15 papers written independently (but amended in the light of joint discussion) by 15 scientists, engineers and economists. Of these 15, a good third posed Weinberg's dilemma; and all of them concluded—some with more regret, some with less—that the cultural component of pure science was beside the point. If an adequate case for the federal support of pure science was to be made out, this must be formulated in terms of the " overheads " doctrine: some ulterior justification must be found for pure science and some index devised for assessing the contribution which it makes, as an overhead or investment, towards other national goals, such as defence, productivity and health.

Now, there is a certain obvious force and attractiveness about this position. It combines hard-headed realism with a far-sighted readiness to take a pioneer's chance, in a way having a ready-made appeal to traditional American attitudes. Furthermore, there is an undoubted justice in the argument that individual business firms and corporations are unlikely (even if they work on the scale of the Bell Telephone Laboratories) to undertake all the basic research work capable of yielding a return in productive terms. Indeed, one of the economists on the N.A.S. panel went out of his way to argue that substantial increases in the proportion of the gross national product devoted to pure research would still continue to pay—in this sense—an " economic return ".[9] At the present stage, it is clear, the scientists from the N.A.S. panel need not fear that the economists would withhold their support for substantially increased federal funds for pure science, even on purely " economic " grounds.

All the same, the case for resting the argument for government support of pure science *entirely* on the " overheads " doctrine—which has been formulated in the following terms:

> the fundamental justification for expending large sums from the federal budget to support basic research is that these expenditures are capital investments in the stock of knowledge which pay off in increased outputs of goods and services that our society strongly desires [10]

—has some embarrassing consequences also. These become apparent, if one inspects the argument more closely. For scientists (regarded as a political group) do not have one single collective aim only—namely, to ensure that the upward curve of total spending on pure science continues to climb as steeply as they " believe to be proper ".[11] In addition, they are committed to another political principle, which is of equal importance to

---

9 Kaysen, Carl, " Federal Support of Basic Research ", in *Basic Research and National Goals* (Washington: U.S. Government Printing Office, 1965), pp. 147–167, especially pp. 151 ff. This paper is reprinted in *Minerva*, this issue, *infra*, pp. 254–272.
10 *Ibid.*, p. 148.
11 Weinberg, Alvin M., *op. cit.*, p. 14.

them, namely, that the manner in which this total amount of support is carved up between different sciences and projects shall remain, so far as possible, to be decided by the scientists themselves. This is one of the few points over which sharp differences of opinion arose on the N.A.S. panel, and it is one on which Dean Don K. Price has commented recently.[12] It prompted the other economist on the panel to comment, sardonically:

> To an important extent, indeed, scientific research has become the secular religion of materialistic society; and it is somewhat paradoxical that a country whose constitution enforces the strict separation of church and state should have contributed so much public money to the establishment and propagation of scientific messianism.[13]

The essence of the resulting problem is this: if the fundamental argument for government support of basic research remains entirely the prospect of economic payoff—even *indirect* economic payoff—then decisions about the subdivision of the total available research funds between different lines of basic research will cease to be a matter for pure scientific judgement alone and become an economic, or even a political, issue.

Now, of course, from the very beginning of the twentieth century debate about this general topic, it has been argued that, in the long run, it was inevitable that he who paid the piper would decide to call the tune.[14] To an onlooker, it is interesting to observe the line of defence thrown up by the scientists on the N.A.S. panel to protect that autonomy, without abandoning the " overheads " doctrine. This defence (presented, for instance, by Dr. George Kistiakowsky at the actual committee hearing,[15] when the report was presented to the House of Representatives Committee) is based on an appeal to the essential *unpredictability* of the technological " fruit " to be expected from investment in basic research: the impossibility of foreseeing, for instance, that work on the microwave spectrum of ammonia would open up a whole new field of maser and laser technology.[16]

---

[12] Price, Don K., " The Established Dissenters ", *Daedalus*, LXLIV, 1, (Winter, 1965), pp. 84–116; *idem, The Scientific Estate* (Cambridge, Mass.: Harvard University Press, 1965).

[13] Footnote 4, p. 141, in Johnson, Harry G., " Federal Support of Basic Research: Some Economic Issues " in *Basic Research and National Goals* (Washington: U.S. Government Printing Office, 1965), pp. 127–141. This paper was reprinted in *Minerva*, III, 4 (Summer, 1965), pp. 500–514.

[14] See, for instance, Bernal, J. D., *The Social Function of Science* (London: Routledge, 1939) and the ensuing discussion, particularly in the publications of the Society for Freedom in Science.

[15] Statement of Dr. George B. Kistiakowsky at Hearings before the House of Representatives Committee on Science and Astronautics. 89th Congress, First Session, No. 3, 25 April, 1965, pp. 8–12.

[16] Pages 170 ff. in Kistiakowsky, George B., " On Federal Support of Basic Research ", *Basic Research and National Goals* (Washington: U.S. Government Printing Office, 1965), pp. 169–188.

This defence needs to be looked at very carefully. For, although there is a great deal of truth in the " unpredictability " doctrine, *so long as one confines oneself to particular cases,* one must think twice before generalising it. Of course, in any one specific case, it will be impossible to say for certain that some new piece of basic research will have significant technological dividends—the lightning strikes where it will. In general terms, and as a matter of probabilities, on the other hand, the " unpredictability " doctrine is surely *untrue*—some areas are much more *lightning-prone* than others. Suppose we compare Townes' work on the maser with (say) some piece of research into Brazilian palaeobotany or radio astronomy: as a matter of *likelihood,* new modes of interaction between radiation and material substances are surely more likely to have practical utility, while the significance of Brazilian palaeobotany or radio astronomy is much more likely to be solely intellectual—throwing light on problems of evolution or cosmology. Incidentally, the so-called " mission-oriented " agencies (such as the National Institutes of Health) take it for granted that the prospective utility of scientific research is not *absolutely* unpredictable but that one can at any rate identify *general areas* of basic research holding out particular promise for their agencies' missions. What else does the category of " mission-oriented basic research " imply?

To go further: there is room for a legitimate suspicion that the " argument from unpredictability " involves an element of naivety, or even of humbug. For, as a matter of practical politics, it has a very useful part to play, as a means of reconciling the statements: (1) that the entire political justification for supporting " free basic research " in general is the prospect of eventual *utility* (the " overheads " doctrine), and (2) that the allocation of this support for free basic research as between the various sciences, and between different research projects in any one science, must be done by scientists themselves on the basis of their own *intellectual* interests (the " autonomy " doctrine). Yet, there is *no* reason to suppose that the scale of likely utility will always coincide with the scale of likely intellectual value. Indeed, there is some reason to suppose that they will often diverge. Before the war, for instance, the Bell Laboratories took Jansky off his early research into radio astronomy, quite properly, at a time when it became reasonably clear that no commercial payoff could be expected from this research. Similarly, at the present time, work on extra-galactic astronomy displays little general prospect of utility—yet its intellectual significance is of the highest.

So, however appealing it may be at first sight, the " overheads " doctrine carries with it very real risks. If looked at coldly, it comes very close to the view of the fine arts propagated officially in the U.S.S.R. under the name of " socialist realism ". I say this, neither in any satirical spirit, nor with any anti-communist intent. Socialist realism, after all, is an *arguable* view of the function of the fine arts, and its forebears go back long before 1917, to Tolstoy, and before him to Plato. Yet, in every form and at every stage, all such attempts to subordinate aesthetic considerations to moral and political ones have carried with them the same implications—in the form of a denial of the artist's right to autonomy in his work. That same denial is implicit in the " overheads " argument for government support of pure science; and appeals to " unpredictability " can serve at best as a delaying action.

The " High Civilisation " Doctrine

The contrary point of view—namely, that government support for pure science must be justified (in part, at least) on *cultural* grounds—was put most recently in the Seaborg panel report to the President. Such support, the Seaborg panel flatly declared, is a necessary part of the price to be paid for maintaining " a high civilisation ".[17] In contradistinction to the " overheads " doctrine, the weaknesses of this position are obvious at first sight. To solicit federal support for pure science by putting it on a level with Bach, Rembrandt and Shakespeare—or, more exactly, with Aaron Copland, Grandma Moses and Robert Lowell—would seem to be, in the present state of things, a Quixotic action:

> The well-paid pure scientists among my friends will undoubtedly object to being converted into scientific Bohemians shivering in poorly heated garrets.[18]

However, in this instance, the case begins to look rather stronger the closer we get to it. As Alvin Weinberg himself immediately goes on to concede, the implied picture of contemporary artists as " shivering Bohemians " owes more to Puccini than it does to the facts of present-day life. One can in fact go further. As things now stand, the fine arts are no longer *frills*. The view of art as something essentially frivolous, decorative and *déraciné* is a late nineteenth-century one, dating from a period when engineers had lost interest in questions of design (" Who cares what it looks like so long as it goes "); whereas, arguably, by now both science and art are essential estates of the realm, in which the " pure " passion for understanding, command and order provides the mainspring of a creative activity which produces both light *and* fruit—just as a tree produces both leaves *and* fruit. Think of the influence of industrial design on the economy of

[17] *Scientific Progress, the University and the Federal Government.* Statement by the President's Science Advisory Committee, prepared by a panel on " Basic Research and Graduate Education " under the chairmanship of Dr. Glenn Seaborg (Washington: The White House, 1960).
[18] Weinberg, Alvin M., *op. cit.*, p. 9.

Scandinavia: as the history of the word "artisan" should remind us, technology is applied *art*, as well as applied science. (The influence of Mondrian on contemporary advertising and fabric design is equally significant.) And, as the technical "features" of industrial products become more and more equivalent, aesthetic "features" will count for more, so that the strength of a country's arts will begin to be of importance, even in the narrowest "economic" terms. Any attempt to depict science as essentially honest, committed, hard-working and fruitful, by contrast with the essentially frivolous, decorative, leisure-time activities of the artist, is therefore doomed to fail, being based on a false antithesis. Both the arts and the sciences are essential parts of any civilised society; and, incidentally, they both have a bearing on economic productivity and profitability.

Evidently, the scientific research of Isaac Newton and Albert Einstein *could not* have been justified, at the time, to a government agency (whether the British Admiralty or the Prussian Ministry of Education) by invoking its marginal propensity to increase industrial productivity, improve defence capacity, or anything like that. Nor, surely, should it really *need* justifying in such terms. Why, then, did the members of the N.A.S. panel shy away with one accord from the cultural arguments for science? The answer to this question lies, I believe, in a challenge thrown out in the course of the panel's proceedings by one of the two economists, Professor Harry G. Johnson of Chicago—a challenge whose force the scientists on the panel felt, but which they could see no way of openly meeting:

> The concept of 'scientific culture' raises a number of questions, among which the most fundamental is the question whether basic scientific research is—in the economist's terms—to be regarded primarily as a consumption or an investment activity. . . . Much of the contemporary 'scientific culture' argument for government support of basic scientific research is such as to put it—intentionally or not—in the class of *economically functionless activity*. The argument that individuals with a talent for such research should be supported by society, for example, differs little from arguments formerly advanced in support of the rights of the owners of landed property to a leisured existence, and is accompanied by a similar assumption of superior social worth of the privileged individuals over common men.[19]

The rest of this paper will be designed to meet that challenge head-on.

The Emergence of Tertiary Industries

If a dilemma such as this is to be resolved at all, it must be resolved boldly. Professor Johnson takes his stand on a distinction between "consumption activities" and "investment activities": he claims that the "science as culture" argument has the effect of putting science "in the

19 Johnson, Harry G., "Federal Support of Basic Research: Some Economic Issues" in *Basic Research and National Goals* (Washington: U.S. Government Printing Office, 1965), p. 132.

class of economically functionless activity ". This position can be satis-
factorily countered in only one way: by demonstrating that the distinction
on which it is based is either *invalid* or *irrelevant*. It may be foolhardy for
a philosopher to intrude into economic theory but that—in effect—is what
I shall now attempt to do.

Professor Johnson defines the purpose of an investment as to " increase
its [society's] future capacity to produce " [20]; however, " culture " is not
to be valued as an investment in this sense; but does it, then, necessarily
follow—we must ask—that " culture " must be economically functionless?
Can economics appraise only activities having an output in the form of a
measurable " product "? Or, alternatively, can we not stand our problem
on its head and demand that the economists should do more to bring
non-productive activities within the ambit of their science?

Let me come to my central point in a roundabout way. Suppose we
look at the economic history of Western society over the last 150 years with
an eye not to input and output but to occupation and employment. A
century and a half ago, the vast mass of the work-force in all countries was
employed in agriculture, forestry, mining and fishing—*primary* industries.
Only during the last 50 years has the proportion of the work-force so
engaged in the major industrialised countries dropped sharply, from well
over 1:2 to something less than 1:10. To begin with, though with periodic
lapses, the labourers who were not needed for *primary* production found
occupation in manufacturing industry and in the associated clerical and
service trades. They did not themselves produce new materials directly
but they did process them into saleable objects—or else they engaged in
other activities (*e.g.*, book-keeping, stock-taking, trucking, cleaning, maintenance) ancillary to such processing. So there came about the age of the
secondary industries.

But, in time, this age too will prove to have been only one phase in
economic history. Even during the great depression, in 1930, J. M. Keynes
could foresee that the increasing efficiency of manufacture—" mechano-
facture " would be a better word—would face us, before the century was
out, with a choice: *either* to employ men in producing unwanted goods, *or*
to find some other range of activities in which to occupy them. Even in
introducing the concept of " technological unemployment ", Keynes was
able to see it, not as a nightmare (the " menace of automation "), but as a
challenge and as a source of hope. The drudgery of production could, in this
way, yield place to more creative occupations.[21]

[20] Johnson, Harry G., *loc. cit.*
[21] Keynes, John Maynard, " Economic Possibilities for Our Grandchildren ", *The Nation
and Athenaeum,* 11 October, 1930, pp. 36–37, and 18 October, 1930, pp. 96–98; reprinted
in Keynes, John Maynard, *Essays in Persuasion* (London: Macmillan, 1931), pp. 358–373.

If this thumbnail picture of recent economic history has any justice in it, then the central economic problem of the coming age will not be one of *output*, but one of *employment*. When Harry Johnson characterised scientists as belonging to " the leisure class ",[22] therefore, his description should not have been evaded—as though it were some kind of a *reproach*—but rather embraced, as a potential virtue.

Indeed, he himself conceded, in small print, that

> In this context ' leisure ' does not mean idleness or frivolous activity, but merely time free from the arduous and uninteresting tasks of producing a subsistence and available for the pursuit of non-material interests.[23]

Pure scientific research is, and can deliberately be chosen to be, one of those new, *tertiary* activities by which employment and prosperity can be maintained in an industrialised society, even after both primary and secondary industries have become *too efficient* to occupy the available labour-force. A scientific research laboratory can serve, just as well as a manufacturing enterprise, as the focus around which the life and prosperity of a community can be organised. To hazard a social prophecy, one can well imagine the " laboratory town " becoming as characteristic a feature of late twentieth-century social life as the " mill town " was in the mid-nineteenth century.

This point can easily be misunderstood, so it must be developed a little further. One virtue of the American space effort is the employment it creates in regions which could otherwise lack a sufficient economic *raison d'être*. Yet, if one presents this argument openly, one sometimes encounters a hostile reaction : even to mention this fact—and it is, surely a fact—strikes some listeners as derogatory. Their minds hark back to Franklin D. Roosevelt's methods of attacking the unemployment created by the Great Depression of 1929–31, such as the Works Progress Administration (W.P.A.), and they exclaim : " Certainly President Johnson does not regard N.A.S.A. as a kind of technological W.P.A.! " People who react in this way evidently still think of the W.P.A. as having been essentially an emergency measure, resorted to in order to deal with a lapse from economic health and normalcy, of a kind that a well-regulated society should be able to do without. For them, non-productive employment still represents a second-best. Yet, at this stage in history, appeals to the nineteenth-century virtues of thrift and hard work are surely anachronistic. We are in a situation in which the maintenance of employment is socially more important than sheer *productivity* alone; and in which the indirect creation of prosperity in a community by the establishment of *tertiary*

[22] Johnson, Harry G., *loc. cit.*
[23] *Ibid.*, p. 141, footnote 3.

industries (*e.g.*, a complex of radio telescopes together with its associated radio-astronomy laboratory) is none the worse, economically speaking, for the fact that the " industry " itself is a *non-productive* industry.

(This is something which J. M. Keynes saw clearly enough on the technical level also. His *General Theory* was, after all, a theory of money, interest *and employment.*[24] And he himself argued that the direct productive output resulting from an investment might be only a small part of its aggregate economic effect: this was " multiplied " by the side-effects produced by the further circulation of the original monetary investment— the fraction of the wages saved and reinvested through local savings banks, etc., etc. By now, experience at places such as Green Bank, West Virginia— site of the United States National Radio Astronomy Observatory—should make it clear to us that Keynes's multiplier effect will continue to operate, even though the direct productive effect of the original investment is actually *zero.*)

Problems of Quantification and Evaluation

Against this background, the dispute between " science as culture " and " science as overheads " no longer has the same force. More and more, in the decades ahead, we shall be free to occupy ourselves with pure science and *happy* to do so. We are, that is, already moving into the world of Maynard Keynes's grandchildren—or rather, great-nephews. This very fact, however, creates some new and urgent intellectual problems for economists; as well as some moral problems for us all.

To begin with the intellectual problems: those economists who have worked in recent years on the economics of research and development, from a conventional economic standpoint, have been faced by problems which were comparatively easy to state, even though they were hard to solve. Suppose that the pounds or dollars which we invest in research and development are intended to lead simply to better, larger or more profitable production, the question then arises: how can one estimate the *degree of efficiency* of such investments? What index or measure can we contrive by which we can establish and compare the incremental changes in productivity resulting from the investment of a further pound or dollar in research and development of different kinds? (This is the theoretical approach adopted in the United States by economists such as Edward F. Denison [25]; and even more so in the case of British economists, such as

24 *Cf.* Keynes, John Maynard, *The General Theory of Employment, Interest and Money* (London: Macmillan, 1936).
25 *Cf.* Denison, Edward F., *The Sources of Economic Growth in the United States and the Alternatives before Us* (New York: Committee for Economic Development, 1962).

Carter and Williams,[26] and Christopher Freeman.[27]) In such inquiries, research and development are seen to be economically functional only insofar as they lead to "economic growth", in the narrow sense of increases in tangible *output*. The resulting questions are not easy to answer but they are at any rate *clear*.

If, however, the central argument of the present paper—about the emergence of "tertiary industries"—is correct, there is a deeper question about the economics of scientific research, which it is hard even to get *stated* correctly. For what I have called "conventional" economics of research and development evaluates the merits of research, only at the price of limiting attention to its side-effects and by-products in the *primary* and *secondary* sectors. (The same is often done in the economics of health, higher education and transportation: there, also, the economic question is formulated in terms of a supposed marginal propensity to increase productivity.) Now, of course, if a particular piece of scientific research *can* be shown to lead to two blades of grass or four tons of copper or 16 automobiles existing where only one existed before, we *can* immediately measure the return on our research investment in monetary terms. The question which economists have not yet squarely faced is, what comparative indices one can use to compare investments in research in situations in which there is no pretence of a conventional "output". Can this be done? If so, how? Those questions are not asked rhetorically.

Perhaps, in one sense of the term, this problem ceases to be an "economic" one as soon as one turns away from questions about productive propensities and so on. Yet, in practice, comparisons between rival research projects must be made and some indices must be devised if research policy is to be carried on in a reasonable way. There is, consequently, a very real danger that the sheer *quantifiability* of productive activities may keep us fixated on problems of increasing the rate and efficiency of output, long after problems of *employment* on a tertiary level (whether in pure science, education, the arts or elsewhere) have become more urgent and practical than the problems of making more consumer goods for less outlay in money and men. And one economist at least, Professor J. K. Galbraith, would certainly resist the limitation of the sphere of "economics" to the primary and secondary industries.[28]

26 *Cf.* Carter, C. F., and Williams, B. R., *Industry and Technical Progress* (London: Oxford University Press, 1957); and subsequent articles: Carter, C. F., "The Distribution of Scientific Effort", *Minerva*, I, 2 (Winter, 1963), pp. 172–181; Williams, B. R., "Research and Economic Growth—What Should We Expect?", *Minerva*, III, 1 (Autumn, 1964), pp. 57–71; and Carter, C. F., and Williams, B. R., "Government Scientific Policy and the Growth of the British Economy", *Minerva*, III, 1 (Autumn, 1964), pp. 114–125.

27 *Cf.* Freeman, Christopher, Poignant, Raymond, and Svennilson, Ingvar, *Science, Economic Growth and Government Policy* (Paris: OECD, 1963).

28 See, for instance, Galbraith, J. K., *The Affluent Society* (Boston: Houghton Mifflin [1958]; London: Hamish Hamilton, 1958).

As always, these problems of economic analysis and quantification have a moral dimension. Those of us who are in a position to observe the attitudes of intelligent young people today, especially in American universities, are very much awake to a strain of austerity and even of asceticism in their attitudes, of a kind which is quite out of tune with national policies based on giving unquestioned priority to problems of productivity and economic growth. The motives behind the success of the Peace Corps, and the unhappiness of those who return, are the attitudes of people who see beyond productivity to something else. As on other occasions in history, the " transfer of allegiance of the intellectuals " (in Crane Brinton's phrase [29]) of which a new arrival in the United States soon becomes aware, is probably a symptom of a major social change. For, behind the technical problem of finding indices for comparing and evaluating tertiary activities, there lies another, deeper problem. On the theoretical level, we need a new economics of welfare, designed to measure not gross national product but gross national benefit: on the moral level, we need to recognise that the aim of a free society is something more than " life, liberty and the pursuit of consumer goods ".

The Social Patterns of Tertiary Society

Perhaps it is unreasonable to demand that economic theorists shall solve these difficult problems *in advance of* events. Perhaps the problems of industrial productivity and output are still too pressing to be ignored, especially in Europe, and in the United Kingdom particularly. Yet there are some hints that, in fact, it is *not* too soon to raise these questions and that even now the events here described are upon us. As so often before changing social patterns may already have begun to take shape, before the academics—or even the politicians—have been able to adapt to them consciously.

The theme of this paper has been, that the pursuit of science is one of those tertiary activities (having no " productive output " in conventional terms) around which our society will be organised in " the post-manufacturing era ". But is this not, to some extent, already happening? Recently, N.A.S.A. had to choose a location for a large new electronics laboratory and published a very full report describing how their decision had been reached.[30] As the appendices associated with this report showed, individuals, Congressmen and chambers of commerce all over the United

---

[29] Brinton, Crane, *The Anatomy of Revolution* (Englewood Cliffs, N.J.: Prentice Hall, 1938; revised edition, Englewood Cliffs, N.J.: Prentice Hall, 1952), especially Chapter 2, section iii.

[30] *NASA Electronics Research Center.* Report to the Committee on Aeronautics and Space Sciences of the United States Senate, 88th Congress, Second Session. (Washington, D.C.: 1964.)

States sent in hundreds of submissions, making out arguments for locating the proposed new electronics laboratory in their own respective communities. By now, indeed, *any* proposal for a large new research installation provokes the same kind of lobbying. This phenomenon was analysed in a recent article in the periodical *International Science and Technology*,[31] on the subject: " How you, too, can attract an R and D industry like Route 128 at Boston, or El Camino Real near San Francisco . . .". What, the writer asked, do you need in order to attract high grade scientists and engineers? Well, sunshine is an advantage, but you need also " art galleries, opera, theatre, museums, libraries, nice places to live, good schools for their kids —and good graduate facilities for themselves—a symphony orchestra, a couple of good chamber music groups, a good French restaurant [etc., etc.] ". [32] In a word, the attempt to attract research laboratories is leading to " the creation of civic environments that are thought to be attractive to technical people ".

> Perhaps we are creating environments that people of all regions have wished for right along, but which did not exist before because of sloth. If this is so, and if we assemble orchestras and build libraries so that our provinces seem less provincial, and if those top-level people still refuse to come—what has been lost? There is still the orchestra. And the books.[33]

Even the *effort* to move into the world of tertiary industries brings with it its own rewards, in the form of a new concern with the quality of life.

Suppose, again, one looks more closely at the " research and development industry " itself. This, too, is changing its character. The early operations of concerns such as Arthur D. Little, Inc., were very much designed with an eye to evolving new manufacturing processes and new products—that is, to " economic growth " in the orthodox sense. By now, however, what began as a symbiosis with productive industry is acquiring an independent life of its own. Research and development firms are taking on government contracts to operate educational services (work camps for high-school drop-outs, experimental teaching techniques for schools, summer institutes in computer science); the basic equipment and stock in trade of such firms is becoming less and less machine tools and inventories and more and more the brain power of their leading men; and, as a result, their turnover has to be measured more and more in terms of intellectual services—regardless of the question whether those intellectual services are directed towards marginal increases in productivity, at one extreme, or the pure increase of knowledge at the other. To those people who are actually

---

[31] Allison, David, " The University and Regional Prosperity ", *International Science and Technology*, (April, 1965), pp. 22–31 and 88–90.
[32] *Ibid.*, p. 31.
[33] *Ibid.*

caught up in the research and development industry, the distinction between the " secondary " and " tertiary " aspects of their work is no longer important. Meanwhile, just as the functional distinction between industry and the university is becoming progressively blurred, so also are the actual structures and buildings of the two types of institution coming to look more alike. With their quadrangles and lawns, their research libraries and seminar rooms, the commercial research laboratories along Route 128 can no longer be easily told, at sight, from their counterparts at colleges and universities.

The starting point of this paper was Alvin Weinberg's argument, that a world in which the pursuit of science for science's sake would be accepted as a necessary part of the whole society's communal activities—as " the business of the community "—is a Shavian dream. On the contrary, I have replied, this world is already becoming a reality. Whether at Menlo Park, California, or Green Bank, West Virginia, or even at Weinberg's own Oak Ridge, Tennessee, the pursuit of pure science is *already* indispensable as a major source of employment and prosperity. To that extent, we have already moved into a " post-productive " or " post-manufacturing " society, in which the focus of whole communities consists in a " tertiary industry "— one whose direct output is not to be measured either in tons or in dollars, or even in marginal contributions to productivity.

In retrospect, that phase of economic history in which all the " economic- ally functional activities " of society had to be concerned with *tangible outputs* will appear to have been a transitory one: as transitory, I suspect, as the corresponding period when the " iron laws of economics " were held to justify the employment of child labour in factories. True, the way past Blake's " dark Satanic mills " has not come from *disdaining* productivity but rather from achieving methods of production and levels of productivity which have begun to release men for better things. This was the world of which John Maynard Keynes wrote 35 years ago in his essay on " Economic Possibilities for our Grandchildren "; and, in that world, scientists will no longer feel any need to apologise for their activities—or to prove that, after all, they are doing their bit to keep the mills grinding, too.

THE WARRANTS FOR BASIC RESEARCH

Simon Rottenberg

Basic Research: Investment or Consumption Good

BASIC scientific research, like every human activity, employs resources that could have been put to some other use. The renunciation of the benefits that might be gained through the alternative uses of these resources is the cost borne by society for having engaged in research. Let those resources be put to use in research, and society has less of other desired objects they might otherwise have produced. If resources were available in such abundance that nothing was foregone as a consequence of research being undertaken—if resources had no costs and were therefore " free "—there would be no need for intellectual defences for scientific activity. Resources are, however, scarce and not free, and there are therefore costs associated with scientific activity. The fact that we inquire into the grounds that warrant scientific research means that we know it is not costless and we therefore search to discover in what sense it is worth society's while to enable basic research to be done.

Two economists of a panel of the United States National Academy of Sciences, Professors Carl Kaysen and Harry Johnson, have offered what they consider to be acceptable warrants from the vantage point of economics [1] and their position has been criticised by Professor Stephen Toulmin, who is an historian and philosopher of science. Professor Toulmin puts forward what he thinks to be superior grounds.[2] All three acknowledge, at least implicitly, that basic research is worthwhile only if there is a benefit or " payoff " to society from using resources in this activity, but they appear to differ about the benefits they seek. Professor Kaysen puts the economists' case thus:

> The fundamental justification for expending large sums . . . [on] basic research is that these expenditures are capital investments in the stock of knowledge which pay off in increased outputs of goods and services . . .[3]

---

[1] Johnson, Harry G., " Federal Support of Basic Research: Some Economic Issues " and Kaysen, Carl, "Federal Support of Basic Research" in *Basic Research and National Goals. A Report to the Committee on Science and Astronautics, U.S. House of Representatives, by the National Academy of Sciences* (Washington: Government Printing Office, 1965), ix+336 pp. Reprinted in *Minerva*, III, 4 (Summer, 1965), pp. 500–514, and IV, 2 (Winter, 1966), pp. 254–272.

[2] Toulmin, Stephen, " The Complexity of Scientific Choice II: Culture, Overheads of Tertiary Industry? ", *Minerva*, IV, 2 (Winter, 1966), pp. 155–169.

[3] Kaysen, Carl, " Federal Support of Basic Research ", *Minerva*, IV, 2 (Winter, 1966), p. 254.

This article from *Minerva*, V, 1 (Autumn, 1966), pp. 30–38.

To which Professor Toulmin retorts:

> Before the end of the twentieth century . . . new patterns of society and employment which are already emerging today will have turned scientific research—even of the most basic, "cultural" kinds—into a "tertiary industry", the social and economic values of which go far beyond all questions of "productive output" . . .[4]

There are here, one may think, the makings of a fruitful dialogue but it does not quite come off and the opportunity is lost largely because Professor Toulmin gives meanings to the economists' somewhat specialised terms which they did not intend them to have. There is, nonetheless, an issue to be joined. Since the economists' terms are capable of being misunderstood, it is best to start with elementary classifications and definitions that will provide a set of categories into which basic research can later be fitted.

Some Elementary Economic Analysis

Resources, including human labour, are to begin with raw, intermediate and final products. They are all things that can be used to make other things and all things that yield, in consumption, what the economists call "utility", for which a conventional layman's equivalent is "satisfaction". All human activity divides into two parts: work and leisure. Both consume resources. Work, or productive activity, yields output (for which another word is "income") which may be in the form of either goods or services. Goods are not easily distinguishable from services in principle but a reasonably proximate distinction can be drawn in terms of the relative durability of the two classes. Goods are more or less durable; automobiles, for example, are more and fresh fish are less durable. Services are instantaneously perishable, although they can have enduring consequences.

Not all production is done at work. A housewife who prepares a meal for her family or tends her child is also producing as is her husband who builds furniture in his workshop for the use of his family. But only all productive activity that is done at work (in the market) and some productive activity that is not paid for (as, for example, the raising of crops for consumption by the farm family) are summed and called output. Conventions accepted by the income statisticians in summing output exclude much that is produced while at leisure. Similarly there is ambiguity in the treatment of consumption. It is ordinarily thought that consumption activities are those that are done in leisure time, but it is clear that if one enjoys work for which one is paid a wage, work is also in part a consumption activity.

[4] Toulmin, Stephen, *op. cit.*, p. 157.

Income (or production or output) is a composite of investment goods and services and consumption goods and services. Investment produces things that are used to produce other things. Consumption goods and services yield utility directly. It is important to see that investment can take the form of services as well as goods. Thus, a mechanic who repairs a lorry, or a doctor who prolongs the life of a man in his productive years, or a teacher who renders a pupil more skilful all perform investment services. Production usually employs both goods (like hammers, nails and lumber) and services (like a week of a carpenter's time and skill) in combination. Goods and services, when used in production, are called " inputs ".

The production of both investment goods and services and consumption goods and services can occur in both the private and the public sectors. It occurs in the private sector if the decision to produce is responsive to the aggregate of expenditure decisions of households and firms expressed in the market-place. It occurs in the public sector if it is responsive to expenditure decisions made by governments or para-governmental bodies.

A machine tool is an investment good; a meal is a consumption good. In the public sector, a dam used to generate power or to irrigate land or to prevent flooding is an investment good; a public recreational park is a public sector consumption good and the lecture of the park guide is a public sector consumption service. Here again there is ambiguity. A playground may be thought to give pleasure to the young, but it may also have the effect of strengthening their bodies; it is a mixed consumption-cum-investment good. Similarly, a monument to a national hero may give pleasure and also assist a community to achieve political stability by fostering and cementing the loyalty of the people to the nation. It is also a mixture of both.

We do not usually cavil about the uses to which leisure time is put. We think it good that all be informed of all the alternatives that are available to them and, indeed, that the number of alternatives confronting each be multiplied. Each of us, too, thinks that some uses of leisure are preferable to others and some of us try to instruct or inform others of what we consider to be superior hierarchies of preference so that they will not use their time "wastefully". But generally we hold to the libertarian rule that each knows for himself what will be most pleasurable for him and, so long as his choices do not damage others, we impose few constraints upon freedom of choice in the disposition of leisure time.

We know the consumption conditions that must be fulfilled if households are to maximise the satisfaction to be derived from the income at their disposal (so that, in principle, we know there can be wasteful consumption), but those conditions are not inconsistent with the proposition that each household should compose its own preference structure. Similarly, we know a set of principles which ought to govern productive behaviour. There are some, and only some, productive activities that are worthwhile for society. These are activities whose outputs are at least equal in value to the whole value of the resources employed to produce them and among those that qualify by this rule, only those are worthwhile in which the last unit of resources used in every productive activity yields a product of equal value in all of these different activities. Whether resources should be used to produce consumption goods and services or investment goods and services turns on the relative values of the two. Given resources should be allocated between these two uses so that the final units of resources put to each use shall have equal values in output.

Investment requires waiting. Resources are withheld from consumption and used to produce an asset which will cause output in the future to be larger by some increment than it would otherwise be. This increment is called an "income stream". Income increments of given magnitudes in the near future have a larger value in the present than such increments in the distant future; the value of an investment is computed by discounting at some appropriate interest rate the income stream it will produce.

Only some investments are worthwhile. The discounted income streams of alternative investments can be reckoned against the cost of the resources used to produce the assets which create the alternative income streams in such a way that rates of return on different investments can be calculated. Only those investments are worthwhile which produce rates of return equal to those of all other investments. If the rate of return on investment in one line is lower than investment in others, investment in that line should be diminished; if the rate of return in one line is higher than in others, investment in that line should be increased.

Socially optimal production and consumption activities will be undertaken if decisions are made in competitive markets (competitive markets are those in which there is freedom of choice and freedom to enter and leave occupations and industries) and if they are responsive to the preferences of households and firms. The only exceptions are those cases in which production and consumption activities produce gains for "third parties" which the producer or consumer cannot possess or which impose costs on third parties of which the producer or consumer takes no account in making his decisions. In the latter cases, it is desirable that collective and political decision-making processes intervene, either by designing rules that constrain or encourage behaviour in the private sector or by causing activities to be undertaken in the public sector.

Economic Analyses and Basic Research

What are the possible defences for basic research that are suggested by this framework? One, clearly, is that of Professors Kaysen and Johnson. If basic research is an investment in knowledge which pays off in the capacity to create a larger output of goods and services from given resources, it might be justified.

We say " might be justified ", rather than " is justified ", because, to know whether it is truly justified, it is still necessary to examine whether the rate of return on this investment exceeds the interest rate and whether the rate of return would not have been higher if the relevant resources had been invested in some other activity, such as, for example, applied research or the replication of something which is already known. Such an examination is fraught with difficulty precisely because it is of the nature of basic research to have no intended output payoff, in terms of goods and services as conventionally measured by the national income statisticians. Where research has an explicit productive purpose we call it " applied research " and it is excluded from the problem discussed here. If there are payoffs from basic research, they are serendipitous and are known about only after the fact.

To know whether basic research should be undertaken or supported *as a productive investment*, therefore, one can only examine past experience and extrapolate to the future. Some of this research has been successful and some has failed. " Success " means that the research has produced discoveries which have been productively applied and have positively affected the output of goods and services. Some of these effects have been more important than others; the " successes " would have to be measured and ranked. The quantity of resources used in research undertakings would be counted as costs to be weighed against the " successes " or benefits.

Failures, too, would be valued by the quantity of resources used in producing them. This would give us an array of net success and failure values which could be arranged in the form of a frequency distribution. This, when plotted, might exhibit the properties of a normal curve of which the modal value may be zero (in which case, the output gains linked to the research, when discounted, would have a value exactly equal to the value of the resources used in the research) or it might be any other quantity. Or the distribution might be skewed either in the direction of success or of failure.

Such calculations done for basic research in different disciplines and subdisciplines would undoubtedly give different distributions. It is not suggested that this kind of calculation can easily be done. What makes it particularly difficult is that it is not clear which particular productive innovations are attributable to a particular discovery yielded by a given basic research project. But if it can be done and if, in principle, it is permissible to assume that the quantitative research/output relationships that existed in the past will also occur in the future, then it should be possible to say at least tentatively whether basic research in general or basic research in some particular fields is worthy of investment.

Let us suppose the calculation was done and it was found that, with regard to output or investment, it would pay society to employ the relevant resources for some other purpose. Are there other grounds on which basic research can be defended? Only one avenue is left to be explored: if research does not confer benefits as a productive activity, the defence for it must be sought in its function as a leisure or consumption activity. Many persons trained in scientific disciplines find gratification in engaging in the search for knowledge "for its own sake". Just as social welfare is served by permitting members of society to expend some of their income for rods and reels and to give some of their time to recreational fishing, it is also served by permitting them to buy laboratory glassware and instruments and to spend their time in research. On the principle that each person is the most appropriate and efficient arbiter of the disposition of his own time, the aggregate of social utility will be maximised by imposing no constraints on consumption preference schedules which include strong preferences for basic scientific research.

Indeed, if the scientist misread the evidence or was excessively optimistic about the probability of expected outcomes and if he undertook the research, at his own cost, as a productive venture, because he believed that it would yield net output gains which would accrue to him, we would still be better off were we to impose no constraining rules. For, though he may waste resources in the process and bankrupt himself, we cannot be certain *ex ante* that he is wrong.

Scientific Research as a Consumption Good

We have referred to cases in which the scientist conducts research at his own expense. But—if the argument for research as a productive investment is rejected—there may still be an argument based on a conception of research as a consumption good for subsidising the scientist at the cost of others.

His costs may be covered by the grants of a patron. In this case, the research provides utility to the patron and can be defended as an acceptable consumption activity by him. This case is not altered even if the nominal patron is a philanthropic and educational foundation established by a person under the terms of an arrangement by which he assigns to others the administration of his funds nor is it changed if the arrangement is to become effective only upon his death. If the person whose assets are used by the foundation chose this mode of expenditure of his resources, it is assumed *a priori* that he derived more utility from reflecting upon the good that would be done than if he had used these sums for some alternative expenditure. The tax laws may have induced him to spend in this way but this is irrelevant. Consumption preferences are fashioned by one's whole cultural experience; tax incentives are simply a component of culture.

There is finally the case of basic research (still unsupported by the test of productive output) whose cost is met out of the public purse. This is an instance in which research is a public consumption good. It would be like other public artifacts such as a fountain, a monument, a museum of art, an opera house, a recreational area, or the changing of guards. A fountain gives pleasure to those who gaze on it. If all fountains were constructed by private persons there would be too few of them, for each would perhaps take account, in the calculus of his decision, only of his own pleasure but he would leave out of the reckoning the pleasure derived by others in seeing the display that he has had constructed. If, therefore, there is to be a number of fountains consistent with the object of maximising the consumption utility of the society, some fountains must be constructed with public funds and the decision to do so must be made through governmental processes.

Basic research which has an insufficient payoff in output, in a probability sense, to justify it and which is done at public expense must seek an analogous defence. Let us imagine a society which finds pleasure in the accumulation of knowledge for its own sake. Let us imagine also that knowledge cannot be kept secret so that if one discovers and adds an increment to the total stock, others come to know of it. If enough of this pleasurable activity is to be ventured, the decisions to seek knowledge must include, in the composition of the data on which such decisions are taken, not only the pleasure of the researcher (or of his private patron) but also that of others who contemplate his work. In the circumstances here defined, these must be collective decisions and the activity must be taken care of in the public sector for some, though not all, research.

We have now exhausted the arguments in defence of basic research. If basic research is undertaken and supported, it must be either because it has positive output consequences, in which case it fulfils the Kaysen-Johnson condition, or because it is a pleasurable consumption activity. If the latter, it may be either a leisure activity of the scientist, or an object of consumption by a private patron, or a public consumption good. Basic research is defensible on any of these grounds.

The defence of basic research as a consumption good is a linguistic variant of the "high civilisation" defence of the Seaborg panel to which Professor Toulmin refers and, indeed, is not excluded by the economists consulted by the National Academy of Science. Professor Kaysen's paper includes the statement:

> The value of basic research has been assessed in terms of other goods, for which it is a necessary input . . . This is a narrow view: scientific research can be viewed as itself a desired end-product . . . it is an aesthetically and morally desirable form of human activity, and the increase in this activity is itself a proper measure of social and national health.[5]

If he says that the "justification" for basic research rests upon the "pay-off in increased output of goods and services," it is because:

> I think it unnecessary to debate the merits of . . . [this aesthetic view], since the investment or instrumental aspects of basic research are in my judgement of sufficient importance to provide a basis for policy judgement independently.[6]

This statement clearly contains the implication that if research is thought to be a bad investment, warrant for it might still be sought as an acceptable consumption good. It is not of course immediately obvious whether the two defences are of equal value or, if we assume that one is superior, what their rank order should be. One may choose to make and prove a case for public support for research as a component of high culture; one would then say, "It might also have value as a capital asset that will in time improve the material condition of mankind—but I shall not examine this since, in my judgement, the public consumption good aspect of basic research is of sufficient importance to provide a basis for policy judgement independently."

If one does this, one is, in principle, on ground at least as strong as Professor Kaysen's and the two might differ only with respect to the rank order of their respective preferences. Of course, the cases on both sides need, beyond bald assertion, the bringing forward of proof. Here, the investment case seems clearly to have the advantage. Past discoveries in pure research have been found to have "spilled over" into a number of industries. (The evidence which might show that the quantitative relationship between research costs and output gains has been such as to warrant the venture is still not quite watertight.) On the other hand, the view which argues for the public support of pure science as a necessary element of a high civilisation implies that society has two alternatives in its treatment of research as a consumption good.

---

[5] Kaysen, Carl, *op. cit.*, p. 259.
[6] *Ibid.*

1. The first alternative consists of only that pure scientific discovery produced by (a) the productive activities of firms which think that basic research is a good investment for them, (b) the individual consumption activity of independent seekers after knowledge at their own cost, and (c) the research of scientists supported by private patrons who find the patronage of science pleasurable *plus* more of other things.

2. The second consists of more pure scientific discovery than would be available under alternative (1) *plus* less of other things.

As things stand at present, society seems to prefer alternative (2) to alternative (1). This preference rests at least in part on the faith that pure scientific discovery is a highly desirable consumption good. It would be worthwhile to design and execute a test that would examine the truth of this proposition, but this has not been made and, until it is made and the truth of the proposition established, it seems best to fall back on the sufficient defence of the investment payoff for which there is plausible if imperfect evidence.

UNDERDEVELOPED SCIENCE
IN UNDERDEVELOPED COUNTRIES

STEVAN DEDIJER

"If you give a man a fish, you feed him for one day, if you
teach him how to fish, you feed him for many days."

Chinese proverb.

I

I AM writing this paper in the hope that it will come to the attention of a select audience of the presidents and prime ministers of those countries where science does not yet exist on any significant scale.

Roughly five out of six prime ministers in the world belong to this group. Today between 15 and 30 of the 120 countries of the world, with less than one-third of its population, possess practically all of its science. They spend more than 95 per cent. of the world's research and development funds in order to produce, first, practically all of the world's research output in the form of research papers, technical reports, discoveries, patents and prototypes of new products and processes, and second, most of the new generation of trained research workers in science and technology. Furthermore, these countries reaped in the past and are now reaping most of the direct economic, political, social and general cultural benefits of scientific research. Finally, during the past twenty years it is mainly these countries which have made the almost simultaneous invention of national research policy as a new institutional mechanism for the development and the use of science to achieve their national objectives.

The other countries—approximately 100 in number—with about two-thirds of the world's population, share in various degrees the remaining one-twentieth of the world's science. They are countries which, either in an absolute or in a relative but very significant sense, have no science.

It has become difficult for these countries to ignore the fact that research is no more than a negligible category in their national division of labour. They cannot avoid being aware that they are essentially pre-research cultures. All kinds of forces, domestic and foreign, political and economic, moral and historical, are acting on the governments of these countries with the inexorability of a law of nature to take some sort of action to promote the development of science in their own countries.

This article from *Minerva*, II, 1 (Autumn, 1963), pp. 61–81.

During these past few years it has come to be realised that under-developed countries are also countries without science. The evidence is presented in Table I.

TABLE 1 [1]

Country	Expenditures on research and development		Consumption of commercially produced energy per capita 1960 (tons equivalent coal)
	% of GNP	$ per capita 1960	
U.S.A. ...	2·8	78·4	8·0
U.S.S.R. ...	2·3	36·4	2·9
U.K. (1961) ...	2·7	35·0	4·9
France ...	2·1	27·0	2·5
Sweden ...	1·6	27·0	3·5
Canada ...	1·2	21·9	5·6
W. Germany	1·6	20·0	3·6
Switzerland...	1·3	20·0	1·9
Netherlands	1·4	13·5	2·8
Norway ...	0·7	10·0	2·7
Luxembourg	0·7	9·3	—
New Zealand	0·6	8·9	2·0
Belgium ...	0·6	7·5	4·1
Japan ...	1·6	6·2	1·3
Hungary ...	1·2	—	2·5
Poland ...	0·9?	5·3?	3·2
Australia ...	0·6	5·3	2·2
Italy ...	0·3?	1·8	1·2
Yugoslavia ...	0·7	1·4	0·9
China ...	—	0·6	0·6
Ghana ...	0·2	0·4	0·1
Lebanon ...	0·1	0·3	0·7
Egypt ...	—	0·3	0·3
Philippines ...	0·1	0·3	0·2
India ...	0·1	0·1	0·1
Pakistan ...	0·1	0·1	0·1

The distress over being a scientifically underdeveloped country is begin-ning to approximate the distress over being an economically underdeveloped one. The growing consensus on the need for simultaneous action on both

[1] Table based on the following data: Dedijer, S., "Measuring the Growth of Science ", Science, CXXXVIII (1962), 3542, pp. 781–788; Kramish, A., "Research and Develop-ment in the Common Market vis-a-vis the U.K., U.S. and U.S.S.R.", P–2742 (Santa Monica: Rand Corporation, 1963). Data on Hungary from a Hungarian government publication.

these problems is reflected in very tangible political actions. The tension between developed and underdeveloped countries is beginning to focus on the scientific gap, in a manner similar to the tension arising from the cleavage between rich and poor countries.

The worldwide discovery that the problem of national development must be coupled to the development of an indigenous science is very recent. Yet, this discovery is now entering into a worldwide consensus.[2] This can be seen, for example, from the activities of many national and international organisations, from the 1,200 papers presented to the UN conference of over 80 countries held on this problem in 1963 in Geneva, from the papers of similar conferences held in Moscow in 1962 and in Rehovoth in 1960, and from numerous papers appearing elsewhere in the world on this subject.

In spite of all the activity and interest in this problem, there is still today a dearth of systematic information and knowledge on some of its basic aspects. Furthermore, since the development of science in countries without it has become a political problem for the advanced countries, political constraints have strongly influenced the mode of its presentation, especially at international conferences. Calling a spade a spade at these conferences has rarely been considered politically advisable either by the participants coming from the pre-research cultures or those from the developed countries.

II

In this paper I intend to deal with those aspects of research policy for countries without science which, in my opinion, either have been stressed insufficiently in the many otherwise excellent papers or have been passed over in silence at international conferences.

One of these questions is what the principal decision-makers of underdeveloped countries should do about science. Decision-making on science in every social system from the highest to the lowest is difficult and it is most difficult when it involves decisions affecting the scientific life of an entire national society. It is also the least studied and the most neglected aspect of national policy even in developed countries. In countries without science, it is very much more difficult, very much less studied and very much more neglected. The future of science in countries at present without it, and its development and use to achieve the objectives of national leadership depend, much more than in any other field of their national policy,

---

[2] Through more intensive contact and communication all countries throughout the world are slowly adopting and pursuing increasingly similar and more compatible national objectives. This seems to me to be a result of greater moral and intellectual consensus and the increased social and political participation of the increasingly educated population in all aspects of the life of the country.

on the interest which they personally take, on their own understanding of it, their own strength of will and their own exertions.

The first effective steps along the road of national development are unthinkable today without using the results of research from the start. It is impossible to estimate your starting degree of development, it is impossible to define your objective, it is impossible to make each step from the first to the second without research in the natural, social and life sciences. An objective estimate of the human and material resources available and necessary for the very first and each subsequent step in development demands the solution by scientists of a series of problems in statistics, demography, sociology, economics, geology, hydrology, geodesy, geography, etc. The efficient development of these resources, whether human, animal or vegetable, demands from the very first a continuous production of scientific knowledge about their specific properties and potentialities. Practically every decision in any field of national endeavour, whether it is the improvement of the trade balance or community development, requires not only know-how but also scientific knowledge produced by research performed in the local environment. Every aspect of national development policy depends on research conducted within the country, although it must, of course, be based on the achievements of, and conform with, the standards of international science. National development requires a large and continuous production of scientific results; the importation of foreign specialists to produce them is politically and economically intolerable as a long-term arrangement. The development of a national research potential, i.e., qualified scientists, scientific institutions and equipment and a scientific culture within those circles must be created in order to carry out other national policies with any degree of effectiveness. The development of this potential must be regarded from the first not as a luxury but as an inseparable part of the general programme of development. Hence, a policy for the development and the use of science must be from the start an integral part of the national policy. Science policy must be as important a part of the national development policy as economic and educational policy and, perhaps, more important than foreign, military and other policies. To neglect a planned and vigorous development of indigenous research in the physical, life and social sciences endangers the whole process of development.[3]

Such is the task. How seriously do the underdeveloped countries take that task? Tables II and III aim to show that the development of science, as

[3] At the Geneva conference this lesson was not so strongly emphasised and not so evident as the first one. The number of papers submitted to the section on science policy was smaller than the number submitted to any other section, amounting only to 2·8 per cent. of all the papers. This shows that the awareness of the importance of science policy is relatively embryonic among scientists and scientific administrators in the developed countries themselves, since the papers for the conference came largely from them.

previously defined was, right up to 1963, both absolutely and relatively, a much smaller public concern in underdeveloped countries than in advanced countries.

In Table II the degree of existence of " science " as an object of concern is roughly estimated for the three countries by the number of papers, articles, books, etc., published in those countries during 1960 on the development of national science. Table III compares the awareness of science for 80 countries which participated in the Geneva conference. In this table, the participation in terms of number of delegates and papers submitted to the conference has been used to compare the degree of concern with scientific policy and the application of science to social development. Though these measurements are extremely rough and apply only very generally, one can see from them that actual effort on behalf of science (as manifested in expenditure on science) and a general extra-governmental awareness of its importance are closely associated with each other.

TABLE II

Country	Population (Yugoslavia = 1·0)	Research effort % GNP spent on R & D	Awareness amount of published discussion of social, cultural and economic importance of science (Pakistan = 1·0)
Pakistan ...	4·5	0·1	1
Yugoslavia ...	1·0	0·7	50
United Kingdom	2·5	2·3	300
All data for 1960			

TABLE III [4]

National income per capita ($) (estimates)	% GNP for R & D	Participation in Geneva conference No. countries	Delegates (per million population)	Papers submitted (per million population)
100	0·1	34	0·14	0·33
200	0·5	18	0·45	0·90
<500	<1·0	17	0·78	1·00
>500	>1·0	15	1·28	1·24

In underdeveloped countries there is less awareness in general public opinion of the importance of science and this is intimately and reciprocally connected with the low priority given to science in development policy and to the carelessness about the cultivation of a scientific potential necessary

---

4 Compiled from United Nations Conference on the Application of Science and Technology for the Benefit of Less Developed Areas, *List of Papers*, E./Conf.39/Inf.3, January, 1963; *Directory of Participants*, E./Conf.39/Inf.7, February, 1963, also *Addendum* 1 and *Addendum* 2 to the *Directory of Participants*.

to produce that science. The farmer, the craftsman, the educator, the civil servant and the politicians in these countries do not see the relevance of science to their concerns. And of course, in underdeveloped countries, there is not a scientific public. There are few scientists or persons who have some measure of scientific education and who follow other professions. There is therefore no representation of the interests of science, no " pressure group " for science, no one to remind the influential section of the population of the need to develop and to apply science. This vicious circle which links the unawareness of the importance of science with its feeble institutional exist-ence and can be broken only by decisions and actions emanating from the central institutional system of the society, from the political elite or from those closely bound up with them and capable of influencing them. There will be no science in the underdeveloped countries unless their political elite become aware of the need for it for their national progress and come sufficiently to appreciate the conditions under which it can be successfully implanted. The first task of the government of such a country, and of its prime minister, is to take the decisions necessary for the creation of scientific institutions, adequately staffed and equipped, and to place them into the right relationships with the educational system, government depart-ments, economic institutions and the organs of public opinion.

Yet, everything so far points to the fact that the prime ministers of countries without science have not yet as a rule (which as every rule has its exceptions) learnt any of the above lessons; they have not even discovered the problem. As a rule, they consider science policy and the development of research work a much less important national problem than their opposite numbers in the advanced countries. One need only compare the program-matic speeches and declarations of prime ministers of countries without science with those of the leaders of such developed countries as the U.S.A. or U.S.S.R. to find out how little attention they have devoted to science and research policy. A similar comparison made between the major political parties and parliamentary bodies of underdeveloped and advanced countries provides similar evidence that there is far too little awareness of the import-ance of science among the political leaders of the former and the institutions which they control. The development of science in the underdeveloped countries needs, together with much else, a matrix of affirmative and informed opinion within and around the political élite.

III

How can this opinion be created where it is now lacking? I would like to make the following suggestion: each prime minister should establish in his office a secretary for science. An intelligent young man with a good under-graduate training in science should be sent to spend a year at the science office of the OECD, at one of the national science policy bodies of the advanced countries (such as the National Science Foundation or Office of the Advisor of Science and Technology in the United States, or at the Science policy seminars at Harvard or Chicago, or, were it possible, to their counter-parts in the Soviet Union, and other countries). He could then be expected to undertake the following tasks: prepare information on problems of the development of science abroad and on the state of science within the country and its problems of growth. He should cooperate with the prime minister's chief of cabinet to ensure that his summary reports on science policy questions constitute an important part of the reading material for the prime minister, that domestic and foreign scientific personalities take more than a negligible part of his time and that questions dealing with science policy are frequently placed on his cabinet's agenda.

Furthermore, steps should be taken to ensure that all members of the cabinet, all branches of government and all leaders of political parties (if more than one) should have access and be exposed to material on the country's scientific policy problems. Even short courses on the importance and modes of interaction of science and society in general and in their own society in particular, held for cabinet members, key parliamentarians and leaders of the key economic sectors would not be amiss.[5]

Journalists and broadcasters should be urged to use their media of communication to arouse the interest of the broader, more or less educated sections of the public on the value of science. It should be arranged that the embassies, high commissions, etc., obtain all publications, reports, etc., concerning science policy originating in the countries to which they are accredited. Copies of these should be sent to the prime minister's secretary for scientific affairs to foster the increase of the social status of the country's scientists and to establish personal and wide contacts with the nascent scientific community.

Those responsible for the industrial and agricultural development of the country should be helped to become aware of the importance of the results of scientific research for the solution of their current problems.

A central research organisation should be established, which will not only have the responsibility of fostering research, but which will also be placed in a position to increase the awareness of the importance of science among all the important decision-making sectors of society, and to establish and support within the universities the academic study of problems of the growth of science within the country.

---

[5] This has been done in some developed countries.

The government of each country should ensure that foreign aid programmes include systematic advice and material help in the establishment and cultivation of scientific research within the country.

Of course all these actions will be of consequence only if the most strenuous exertions are made to promote the conduct of creative scientific research. Scientific teaching and research in the university must be accorded the highest priority. Financial support, administrative facilities, freedom from red tape in the importation of equipment must be provided. Special care must be taken within universities and research organisations to see that creative research is not sacrificed to administrative protocol, to see that young scientists fresh from their advanced degree research are not sacrificed to older men whose interests are largely administrative. Great care will have to be taken to see that the small scientific community and its even smaller subdivisions are not allowed to wither away in isolation, to lose contact with each other, with their peers in neighbouring countries and with their colleagues, teachers and former fellow-students in the wider international scientific community.

IV

" The soil of X is deeply inimical to the growth of science ", a scientist from a highly developed country recently stated in a private letter after visiting an underdeveloped country known for its prime minister's awareness of the importance of science. His statement expresses what every scientist or scientific administrator I have met has said about the situation in underdeveloped countries. I have repeatedly met scientists from underdeveloped countries, who have made such statements as: " My country wants neither me nor any other scientist ", or " The government of my country is inimical to science ", etc. The reluctance of the highly trained young scientist from an underdeveloped country to return to his own country upon completion of his training is not simply attributable to deficient patriotism or enslavement to the money bags and flesh-pots of the advanced countries. In many cases it is motivated, at least in part, and in some cases it is entirely motivated by the knowledge that it is difficult to do good research in their own countries. Not only is equipment and financial provision incomparably poorer than it is in the advanced countries, but scientific administration is usually far more bureaucratic and antipathetic to the needs of scientists for freedom from petty controls. Moreover at home, scientists are few and isolated; there are too few for stimulating interaction and there is none of the atmosphere of excitement which arouses curiosity.

The sources—institutional, cultural, psychological, economic and political—which give rise to such widespread beliefs about the uncongeniality of the condition of the underdeveloped countries must be understood and dealt with if science is to grow there. Although widespread public understanding is desirable, it is especially urgent that the head of government of such a country understands the necessities of a scientific policy, for he has at his disposal the only institution and the only resources that can initiate and support the institutions necessary for the growth of science.

The cultural obstacles to the growth of science in the underdeveloped countries come from a plurality of sources, traditional and modern, indigenous and exogenous. To begin with there was no such thing as modern systematic, theoretically oriented science or scientifically based technology in the traditional indigenous cultures, even of those countries inheriting the great world religions. There was some empirical medicine, some astronomy and mathematics, but little else. Certainly what we call the scientific outlook, the belief in the value of systematic and persistent observation as a means of discovering the coherence and determinateness of the natural order of existence, was not widely diffused in any of these traditional cultures. Such indigenous science as had once existed has long since died away and no traces of it are still active in the contemporary form of these traditional cultures. In the modern sector of the cultures science has not played a large part. The educational system, beside the fact that it left most of the population untouched, had very little scientific content. The universities were primarily literary and abstract in their orientation, the civil services were modelled on the metropolitan services which stressed humanistic, legal and administrative studies (occasionally mathematics) as preparation for entry, and politics naturally had no place for science—even radical and socialist politics which spoke of planning and of "scientific socialism". This was approximately the cultural situation of science on the accession of independence and it has not changed greatly since then.

Curiosity, the pleasure of discovery, the readiness to discard previously held views in the face of new observations and new theories have not begun to spring from minds still rooted in the indigenous traditional and colonial modern cultures. The institutions which generate such dispositions and which keep them alive are difficult to create. They involve an intricate structure of relations within departments, of departments within faculties, faculties within universities, universities in relation to public authority, governmental research institutions and many others. These are only the most external aspects of the system. More fundamental are the normative and motivational systems, the models of action in past and present, which underlie and permeate the social relationships. In any case, they scarcely exist at present in underdeveloped countries.

Underdeveloped countries are pre-research cultures lacking the institutional and motivational elements. Hence, they are basically alien or hostile to almost every aspect of research and the utilisation of its results. The embryonic science developing in this environment will show in every one of its cells, that underdeveloped countries have underdeveloped decision-makers on science, underdeveloped research councils and science advisers, under-developed administrators of science and underdeveloped scientists. It is not that scientists in underdeveloped countries are technically untrained or technically incompetent; it is rather that, being a part of their national culture, they will themselves lack, or will not be able to impose or recreate in their society and culture, so alien to science, those fundamental orientations (if they have them) which are necessary for really productive research.

It is under such conditions that political leaders will have to promulgate policies and execute decisions on science. They will be without the support of these three major, more or less autonomous sectors of society which, in the advanced countries, contribute so much to the growth of science:

(1) A scientific community with its own institutions of training, research and communication and its own scientific tradition.

(2) A government apparatus—politicians and civil servants—with a tradition in dealing with science, making some provision for it or at least appreciating its intellectual and practical value.

(3) Industrial, agricultural, commercial, educational, medical, military and other institutions, which have learnt the value of the results of research and have learnt to make more or less reasonable demands for them on the scientific community and the government.

There is another constraint further complicating the conditions under which the political leaders of underdeveloped countries must work in general and on science in particular. Foreign pressures and domestic centrifugal forces give rise to grave political, economic and social strains and insta-bilities in every underdeveloped country. In those where the prime ministers are endeavouring to modernise, these strains and instabilities are incom-parably greater. The political elite are especially prone to experience these strains which threaten all their past and future achievements. Their response to these threats is to block free communication, with the result that they fail to learn what science can offer them, and they also render the conditions in their countries even more unattractive to their own and expatriate scientists.

So we see that, in dealing with science, a series of pitfalls and traps are set for the political elites of underdeveloped countries. Their own lack of knowledge, experience, the underdeveloped cultural environment of which they are parts and which influence the formulation and execution of their decisions, and their own sensitivity to real and imagined threats to the stability of their countries, all help to obscure their vision.

Because of these factors the political elites of underdeveloped countries will be even more prone to make mistakes about science than their colleagues in the developed ones, no matter how able they are personally or how much they are aware of the general importance of science. Even when they try to make research activity a part of the national division of labour, they can foster a series of malformations in their scientific institutions, which might take a long time to diagnose and cure.

V

The present supermarket of world science contains so many attractive goods so expensive for the purse of the underdeveloped country, that only through judicious budgeting of their resources can they buy those most suitable for their capacities and useful for their particular practical and scientific purposes. Every decision on science because of the nature of research and our ignorance is liable to be much more complicated, much more uncertain and hazardous than almost any other type of decision. Even in countries with the most developed science, decisions about it are bound to be leaps into the dark by more or less blind men. Then, of course, in addition to the sheer element of unpredictability inherent in scientific research, many other factors enter into decisions in scientific policy. Ideology, concerns of political and economic advantage, even the pressure group-like action of particular sectors of the scientific community can influence decisions in scientific policy. In underdeveloped countries, ignorance, prejudice and the absence of sources of reasonable advice render such decisions much more difficult, their success much more problematic. There, because of the lack of check and balance within and by the scientific community and other institutions, e.g., industrial and military, with some experience in dealing with science and because of the lack of " personal ", if not technical knowledge, of the nature of research work, decision-making on science is liable to go even more astray than in advanced countries.

The small number of scientists, the gerontocratic tendencies of the traditional culture, the hierarchical civil service traditions of the modern sector of the society, the concern for national prestige and for "monuments", the preoccupation with metropolitan models, plain fraudulence among many of those who offer themselves as advisers, are only a few of the many impediments to rational science policies. In underdeveloped countries, powerful and misinformed military, economic or political interests, native scientists with real or fictitious scientific achievements and no experience in science management are capable of sending science expenditures down the drain for years on completely unrealistic projects, simply because the decisions were made *ad hoc*, without an open discussion on the basis of widely gathered information and advice from home and abroad.

To reduce the probability of such outcomes every decision on science must be part of a national plan for the development and use of the results of research. Science must be looked upon as part of a planned national policy. The formulation and control of the execution of research policy and its continuous improvement has to be one of the constant major tasks of the highest political leader of an underdeveloped country. This means that he must participate in the basic decisions on all of the key components of a national science policy. This means that he must take personal responsibility for the principal national objectives to be reached by the existing research potential of the country; the planning of the growth of various elements of this research potential, such as scientific and technological manpower, training and research institutions, equipment and buildings, scientific publications, financial support; long-range research programmes; the distribution of scientific potential with respect to the prospective productive forces of the country and with respect to the principal goals to be reached through science. This calls for decisions on priorities in the establishment of institutes for such basic survey services as geodesy, meteorology, demography, geology, pedology, hydrology, etc.; for institutes in basic and applied research and development in natural, social and life sciences; research institutes for work in special fields like nuclear energy, automation, military research, etc.

The political leader of an underdeveloped country must also see to it that distribution of the research effort and the research potential between the government, the economy and the universities is an optimal one which serves the needs of the former and does no harm to the latter. He must also attend to the phasing of the scientific effort and of the utilisation of the scientific potential. In other words, he must give personal attention to the drafting of a plan for the development of science and its continuous revision in the light of past mistakes and emergent tasks.

Finally, the president or prime minister must initiate or support a series of decisions on measures which aim at making research productive of intellectual and technological results. To be able to accomplish this tremendous task, it will be incumbent on the political leader to remember that there is no such thing as spending too much on research and development.

VI

Investments in science, though giving abundant returns of all kinds including money, are not " get-rich-quick " investments. The hope, most often unrealised, for quick and bountiful returns from investments in apparently more urgent undertakings gives rise to a situation in which research expenditures are pushed lower and lower on the list of investment priorities of the underdeveloped countries. As may be seen on Table I, the more underdeveloped a country, the smaller the proportion of its income invested in scientific research. That is why, as I pointed out several years ago,[6] the rate of development of science in underdeveloped countries probably is lower than the growth of their economy. As a result, the under-developed countries condemn themselves to lag further and further behind in the growth of science and the utilisation of its results. This means that they can expect to be beaten on the competitive world market even in the future by countries which invest more in research. A twofold increase in research expenditure every two years for the next decade will barely permit an underdeveloped country to keep up with the growth of investment in research in the developed countries. If he decides, as he should, on such a course, a president or prime minister of an underdeveloped country will have a hard time with his finance and other ministers. He must, however, persist, keeping his eyes on the future and pointing out that many, if not most, of the " get-rich-quick " investments in industry, agriculture, power, mining, etc., have not realised the hopes which were placed in them.

Just as he must insist with unremitting obstinacy on expanding the allocations for science and technology, so too he must insist that his country cannot have too many scientists. Underdeveloped countries have so few scientists, engineers and doctors per million population, compared with the developed ones, that however great an expansion they undertake, they will still lag far behind the more advanced countries for a long time to come. Furthermore, as it has been shown, the number of science and technology students per million population is at present about a tenth as large in countries with an income of $100 per head as it is in those of $500 or more. Endeavours to correct this balance must constitute one of the central, if not the central, science policy tasks of political leaders.

In the countries with an income of $300 and less per head, for which such data are obtainable, i.e. about 15, one notes that in all of them without exception, what research does exist in the country is isolated from the universities, where students can specialise in science. As a matter of fact, one could almost define an underdeveloped country as a country in which all research is performed, not in industrial laboratories or in the universities, but in large, almost hermetically sealed, government institutes. The arguments for the relatively small number of science, technology and medical students and for this divorce of research from training in scientific

6 Dedijer, S., " Scientific Research and Development : A Comparative Study ", *Nature*, CLXXXVII (1960), 4736, pp. 458–461.

research work have been repeated so often that they have become almost dogmas. Political leaders must demand a careful review of this situation. There is a certain blindness to the importance of scientific manpower in underdeveloped countries. It therefore seems acceptable to isolate so much of the training capacity of the country, however meagre that capacity may be, from the prospective trainees. The reasons for this neglect are obscure. Perhaps it is because training has come to be thought of as a postgraduate activity into which the universities of underdeveloped countries have not felt it incumbent on themselves to enter, partly because the political and journalistic elites have seldom been postgraduate students themselves. But whatever the cause, manpower plans, even those which devote much attention to engineering, medicine and teaching, do not pay much attention to the output of scientists and even less to the output of scientists who can train more scientists.

The determination of the proper balance in a national research programme in basic and applied physical, social and life sciences represents one of the most difficult research policy problems for the underdeveloped countries. Yet, there is a simple rule which can guide the decision-maker in facing this task.

The need to survey the human, mineral and vegetable resources of the country should constitute the first claim and should determine what kind of research institutes the country should invest in first of all. A country with abundant agricultural resources of a given kind and with the prospect of developing them, would do best to concentrate intense research efforts in all branches of science, life, natural and social, basic and applied, which would increase the knowledge bearing on agriculture. The development of basic research in these circumstances is indispensable from many standpoints; it is necessary for broadening the range of practical possibilities available in the given complex of resources: it is necessary for the maintenance of research standards and research morale, etc.

The survey and development of resources should definitely include the development of human resources. For this reason, a vigorous development of the social sciences must not be neglected. The social sciences in the developed countries, both East and West, play an ever increasing and indispensable role in the formulation of national policies and in the evaluation of their execution. In underdeveloped countries, the social sciences have the special task of bringing the mirror to the face of the nation to show what it is: an underdeveloped culture, and to illuminate, as much as its present methods permit it, the rough road of the cultural revolution, which is necessary if the country is to develop.[7]

---

[7] The social sciences can contribute not only inventories of what the country possesses, they can also increase and enrich the national self-consciousness, the mutual awareness

The political leadership of an underdeveloped country, even if it wishes to develop science as vigorously as possible, must confront not only its own ignorance of the problem but also the difficulty of getting informed and disinterested advice. Eagerness to develop science, ignorance, numerous distracting preoccupations all make it easy for cranks and promoters to come forward and to acquire influence over scientific policy.

VII

Certain of the malaises of the scientific life of the underdeveloped countries are well known and we have already referred to them. But there has not yet been sufficient analysis of them. For example it is frequently, and on the whole correctly, asserted that a tradition of scientific research is lacking in underdeveloped countries. We have, however, still to understand what is implied by this observation which is of the greatest importance. It is a matter of attitudes towards discovery, capacities to perceive problems, to sense the relevance of theories to observations, and vice versa. It is in part a frame of mind, which emanates from one person, which is received, perhaps through identification, by another person and which is constantly renewed and revised. The conditions of this transference of a general outlook, the internal arrangements within an institution, the optimal combination of different generations—on such matters we as yet know very little. Yet if scientific policy is to be made effectively it must take into account the necessity of implanting and continuously reproducing the tradition of science, that paradoxical combination of continuity and innovation which is at the centre of scientific growth.

How to create something important which is lacking is certainly difficult. It should not stand in the way of clearing the ground of obvious malformations, the presence of which can only impede the growth of science and which affect almost every detail of the formulation and execution of national science policy. Such malformations, some of which will be referred to later for illustrative purposes, affect the work of the individual research worker, of the directors of laboratories and the highest scientific administrators in the country. They are expressed in the national research programme and in the outlook of research workers and in their relationship with each other. Those which I mention are only a few of the many which exist; all are worthy of detailed analysis because, unless they are well

of the different sections of the population and thus aid in the transformation of a collection of scarcely integrated separate traditional societies united by a single government into a coherent modern society. This in turn, by unifying the culture of the country, would help to feed the reservoir of talent from which prospective scientists could be drawn.

understood, their pernicious character will not be appreciated and their elimination will be delayed.

There often exists a belief that the development of a particular branch of science, or of a particular project, will enable the country " to jump generations ", to turn it overnight from an underdeveloped to a developed culture, or to make it militarily vastly superior to its enemies. Although I have encountered only two leaders of underdeveloped countries making such statements openly and repeatedly, I conclude from many interviews with scientific policy officials and from the policies themselves that the belief is fairly widespread.

Another malady of science policy is to concentrate most of the national research effort on one major project and to plan its development far beyond the actual capacities or the need of the country for this particular branch of research or its results.

The tendency to devote resources to research programmes which confer prestige internationally is another vice of contemporary science policy in the countries we are considering, much more so than in the developed countries.

Even countries with the minimal provision for science are bound to work through a number of institutions, agencies, departments, etc., which compete for the meagre funds available from the science budget. Very often these agencies are concerned with particular projects arising out of the real (or imagined) needs of the country. They are often ruled by powerful personalities, who are imbued with a missionary spirit on behalf of their particular branch of science, and who act as " scientific empire builders ". Each of them rules his " empire " with an iron bureaucratic hand, keeping most of his staff in misery and eager to leave the institution and even the country at the first opportunity. Sometimes the aspiring " scientific imperialists " make pacts with other scientific empire builders not to recruit staff from each other, thus reducing mobility and demoralising their " captive scientists ". This feature is accentuated by the civil service-like bureaucratisation of scientific institutions in the underdeveloped countries, which, among other rigidities, prevent scientists from attempting to transfer to another research institution unless they have the permission of their present superiors.

More generally, civil service procedures result in almost unbelievably frequent and changing demands on scientific institutions and individual scientists for " proper procedure "; there is a distracting insistence on the control of everything and everybody connected with research and all this requires " reports ", " plans " and " requests in proper form ". In the more advanced countries there has been sufficient experience with scientists and there are enough scientists in scientific administration to inhibit such bureaucratic *paperasserie*. In the underdeveloped countries, there has been too little experience of scientific research to permit a realistic image of how scientists really work and of what is compatible with their effectiveness. There is not enough of a culture of the scientific community to stand up against the tradition of the civil service and the distrustfulness of politicians.

Bureaucracy, beliefs in the magical powers of science and empire-building are usually accompanied by an insistence on secrecy. The jealous insistence on secrecy in underdeveloped countries is very often used as a means to protect scientific cranks and rogues who proliferate in countries without scientific tradition.

In the lower ranks of the small and fragmentary scientific community of the underdeveloped countries, there is a pronounced tendency to build " institutes " as private reserves, from which to attack and repel, as if from medieval castles, particular academic and scientific enemies.

All decision-making in government entails a political element and the higher the level of decision, the more pronounced is this political element and the more pronounced is the struggle to exercise influence and power. This is true of decision-making in science and science policy in every country. In underdeveloped countries these struggles for power sometimes become so acute that often the basic objectives of decision-making are lost in an atmosphere of political wire-pulling, lobbying and acrimony.

These vices are not, as I have already suggested, confined to the uppermost levels of science policy-making; they intrude into the " atmosphere " of laboratories, of the whole institutes, of branches of science and of the whole institutional system of science in some countries. They help to accentuate the attenuated creativity of diminution of the scientific workers of these countries.

VIII

There is a small number of first-class scientists in some underdeveloped countries, who, under the extremely difficult conditions existing prior to independence, when scientific research was scarcely encouraged, managed not only to survive scientifically but, through the strength of their character and their devotion to science, also managed to build first-class laboratories and institutes. The hard struggle for scientific existence they had to go through often left an imprint on their personalities, which made it difficult for some of them to adjust themselves to the new generation of scientists. Such personalities, however, as a rule hold very high the standards of fruitful scientific work and they often succeed in transmitting their views and beliefs to their pupils.

Yet, one often encounters other types of scientific personalities who have a more negative influence on scientific productivity. Some of these like " Diderot's monk " do as little work as possible, always speak well of their superiors, and let the rest of the world, including their fellow scientist monks, go to the devil. There are also " research politicians " who devote little time to their own research work but much time to laboratory politics, flattering their superiors, jockeying for posts and stipends for travel abroad.

Then there are the irreconcilable malcontents who complain incessantly about the conditions of their laboratories, the condition of science in their country, their colleagues' incompetence but do very little themselves.

Other frequently encountered types of scientists who help to hold science back in the underdeveloped countries are the " cranks " who have fixed ideas about solving one or more unsolved problems in science by methods which border on lunacy. With powerful political backing these cranks sometimes prosper for a long time, especially if they are, as is not rarely the case, " rogues ". These are sometimes men of ability but they do not do research; their aim is to establish by hook or by crook " a reputation " out of all proportion to their real value as scientists.

All these corruptions of the scientific career can prosper, and have prospered, in advanced countries and in highly developed scientific communities. There, however, sooner or later, through the action of an open scientific community, they are more liable to exposure; in any case they very seldom attain any marked influence. In underdeveloped countries, however, such persons can become dominant in research institutions and can do much damage to the more gifted and more devoted individual scientists as well as to the scientific system as a whole.

These are all contributory to the lesser effectiveness of scientific effort in the underdeveloped countries. A wise president or prime minister will see to it that his adviser for scientific matters initiates studies which, although they are difficult, would measure and assess the research attainments of individual scientists, of whole institutes, the extent of utilisation of research results in industry, agriculture, medicine, etc., in his own country and compare it with similar indicators for other countries. Such inquiries would make more vivid and convincing the general impressions which are gained by less systematic observations concerning the harmful repercussions of " the lack of scientific tradition ".

IX

There are three main causes for these obstructions to the effective use of the limited scientific resources of the underdeveloped countries. Each is interconnected with the other and together render difficult an efficient science policy. They are (1) the failure of such scientists as there are to carry and transmit the tradition of science and their consequent failure to constitute a scientific community; (2) the lack of appreciation of and demand for research by industry, agriculture and other sectors of society; (3) the government, a bureaucratic institution, lacking in experience and tradition in dealing with science is, and must be at the beginning, the only decision-maker on science.

But since the government alone has the funds and, to the extent that they exist, the powers to promote the development of science, it is up to the government and to its president or prime minister to remedy this situation. This can be done if the government, at least in its crucial parts, can change its character, and that being done, the first of two factors are made the major and most important concern of the research policy of the country:

(1) A scientific policy, to be effective, must build a scientific community with its own traditions, closely linked to the international scientific community and the universal standards of science.

(2) To foster the appreciation of and demand for research work and research results by the industry, agriculture and by the other major institutional sub-systems of the society.

The second of these objectives is a very complex problem depending on the economic and all the other national policies. Creating demand for science in industry, instilling in industry the need for inventiveness, the need to "make better mouse-traps" is a very difficult problem, which nevertheless must become a major objective of national policy. This is in general one of the key problems in most countries with highly and rigidly planned economies—economies in which no substitute has been found for the profit motive and the competitive effect of the internal or foreign market is not operative. What to substitute for the profit motive and competition as the drive for inventiveness in the economic life, which has to be increasingly based on continuous and long-range planning, is a problem which not only the underdeveloped, but the communist countries too, are beginning to face. The study of this problem and finding ways to stimulate the demand for research in agriculture, industry and all other economic sectors must be a special field of concern for those making national scientific policy. This is especially true for underdeveloped countries, where research institutes in applied science are often in fact " pure " research institutes, simply because their results are never demanded or used by their native agriculture or industry.

To attempt to solve the problem of making a scientific community with a scientific tradition, the political leaders of an underdeveloped country will have to turn to the social sciences for basic information, for particular studies and for empirical understanding on how to proceed.

Underdeveloped countries have scientists, but have no scientific community, if we define the latter, in the light of what exists in the developed countries, as an organised group with a developed system of beliefs, with a developed system of institutions for internal communication, as well as a system of communication for dealing with other social groups, and which is bound by certain traditional norms of behaviour for furthering their individual and collective work in science.

In an advanced country, the scientific community is large enough to permit differentiation with sufficient members in each special field to permit complex interaction with each other, and sufficiently different from each other to be able to stimulate each other; it must be organised into a system of training, research, publicity, standard-protecting institutions, which control admission by rigorous scrutiny of qualifications, and which represent and protect the community in its relations with other sectors of the society. It has its own culture, i.e. beliefs, values and vocabulary, its own heroes and models of conduct; it is aware of its history and its continuity with the past without being rigidly attached to that past. It has its own system of communicating and assessing the results of research and analysis and its own patterns of promoting the proficient and holding back those who do not meet its requirements. It has its own circles of face to face interaction and inter-individual communication. It is aware of itself at least in a vague way as something different from other sectors of the society, and it is recognised and acknowledged as a distinctive group by those other sectors of the society. It is linked with other scientific communities across political boundaries by personal contact, by mutual appreciation and by public communication and formal association as well as by a sense of fundamental affinity. It is aware of its identity within the national and world societies and has pronounced beliefs in the legitimacy of its role and calling. We can see how far the underdeveloped countries are from having a scientific community in this sense. Still that must be the objective of their policy.

In the underdeveloped countries scientists are relatively few in number, and they are often, as far as any particular field of research is concerned, dispersed over long distances. They suffer from isolation from each other and thus they do not have the benefits of the stimulation of the presence of persons working in closely related fields. They are in danger, a danger to which they too often succumb, of losing contacts with their colleagues in the international scientific community. They feel peripheral and out of touch with the important developments in science unless they can visit and be visited by important scientists from the more developed countries; they feel inferior and neglected because their own journals and organs of publication, where they exist at all, are seldom read by foreign scientists, seldom quoted in the literature and are indeed often neglected by their own colleagues at home. They have little contact with their colleagues in neighbouring underdeveloped countries.

They are in brief not fully-fledged members of the scientific community and their work suffers accordingly. Neither its scientific nor its practical value is what it could be, given the talent and training of many of the persons following scientific careers in underdeveloped countries. And the scientific community in the country being frail and anaemic, its relationships with the political order of its own country is poor too. Even were the political leaders willing to pay attention to it, its advice would be made unrealistic by the isolation of its advisers from the pulsations of the living scientific community. The strengthening of the internal structure of the scientific community of an underdeveloped country would improve the quality of its work, enhance its selfconfidence and increase the weight which it carries with the other sectors of its society.

This leads us back to the point made above. The fruitful pursuit of scientific truth and its application, once discovered, is not just a matter of talented individuals, well trained in foreign universities and supplied with the equipment they desire. These are very important, but the cultivation of science is a collective understanding and success in it depends on an appropriate social structure. This social structure is the scientific community and its specialised institutions. The underdeveloped countries can dispense with it no more than the developed countries can. The advantage of the latter is that they have inherited it and can develop it as the occasion demands, with the effort which is inherent in traditions which are already well established. The disadvantage of the underdeveloped countries is that they must develop it *ab ovo*. But develop it they must or they will have no scientific development and no economic and social development either.[8] Until science becomes an autonomously growing institution in the new states, all devolves on policy.

[8] The identification of the standards of judgment and conduct of scientists as a community is a very difficult task. Such norms are transmitted, mostly orally, through example. They are not easily articulated. Very rarely could one find examples of them described in the biographies or autobiographies of scientists. Nonetheless, the social sciences are now beginning, however rudimentarily, to analyse them. Progress is being made in the perception and articulation of the nature and content of these standards, and the community which is maintained by them and by which they are maintained. Even now social scientists are beginning to think in the social science idiom. Thus, for example, we find in *Science*, CXXXIX (1963), 3561, an editorial on " The Roots of Scientific Integrity ", written practically in the language of sociology, where among other things it is said: " Part of the strength of science is that it has tended to attract individuals who love knowledge and the creation of it. Just as important to the integrity of science have been the unwritten rules of the game. These provide recognition and approbation for work which is imaginative and accurate, and apathy or criticism for the trivial and inaccurate. . . . . Thus, it is the communication process which is at the core of the vitality and integrity of science. . . . The system of rewards and punishments tends to make honest, vigorous, conscientious, hardworking scholars out of people who have human tendencies of slothfulness and no more rectitude than the law requires ". The social sciences are, however, only in their beginnings in this subject. The natural science policy-makers must draw from them what help they can in getting on with the task of supporting, fostering and encouraging the creation and development of their own scientific communities.

TECHNICAL ASSISTANCE AND FUNDAMENTAL RESEARCH IN UNDERDEVELOPED COUNTRIES

Michael J. Moravcsik

I

Technical assistance programmes to the developing countries have been in existence for some years now and they are carried out on a considerable scale. A large majority of such programmes consist of applied research and development in connection with a specific problem, such as the survey and the proposal for the solution of the salinity and waterlogging problems in Pakistan. In such programmes, " experts " on the particular problem are selected to evaluate the situation and design the solution, and then financial assistance is given to the country involved to carry out the recommendations in cooperation with or under the supervision of the " experts ". Engineers, applied scientists, legal experts, administrators, expert farmers, industrial managers, etc., are the chief participants in such undertakings. These programmes are usually of fairly short range (meaning that they affect only the next two or three five-year plans) and can produce strikingly tangible results, which contribute to the well-being of the country involved.

Another kind of assistance is concerned with education. Considerable American foreign aid funds are spent in various countries for elementary school building, on demonstration equipment in various colleges, or even on school milk programmes (which, by promising a cup of milk to the student, reduce absenteeism in schools in some cases to a surprisingly large extent). Furthermore, the Fulbright and Smith-Mundt programmes send college and university teachers to various countries to raise the level of instruction and to supplement teaching staffs which are often deficient in number as well as in quality.

There is, however, another aspect of technical assistance which so far has been neglected, which is of major importance to the developing countries and which could be undertaken with a minimum expenditure of funds and a certain amount of cooperation by the scientific communities of the United States or other countries with technical aid programmes. This is fundamental scientific research in the underdeveloped countries. By fundamental research I mean investigations undertaken in the natural sciences for the purpose of enlarging the general body of knowledge

This article from *Minerva*, II, 2 (Winter, 1964), pp. 197–209.

and not because the results to be obtained appear to be helpful in solving a specific practical problem. (Fundamental research in this sense is also sometimes called pure science, as opposed to applied science.)

II

It is not at all obvious that the underdeveloped countries are badly in need of fundamental scientific research today, at this very rudimentary stage of their development. Yet, the need for this kind of research is basic to the argument proposed in this article. If one travels in any of these countries and observes the high rate of illiteracy, the daily struggle to reach the bare subsistence level and the snail's pace at which the modernisation of the social structure moves, one might gain the impression that what these countries need is intense assistance to solve the most urgent economic and administrative problems and that pure research is an unjustifiable luxury. A somewhat more searching investigation will however support the contention that the underdeveloped countries must begin without delay to develop their scientific resources in the direction of fundamental research.

It is of course seldom questioned that applied research is of great importance to the underdeveloped countries even today. There is a great deal of applied research which is done in the advanced countries and which produces results which are relevant to the problems of the less developed countries. This would appear to reduce the urgency of doing even applied research in the underdeveloped countries; after all, what they must do, it could be argued, is to import and apply the results of the research done in the advanced countries. Yet, by its very nature, applied research often tends to be specialised, and conversely, particular problems require separate efforts by applied research workers. It inevitably happens, therefore, that each country finds itself faced with certain problems which must be dealt with scientifically but which are peculiar to it, perhaps because of the climate, the fauna, the natural resources, the economy, the cultural heritage or the geographical location. No advanced country can then be expected to deal with these problems and only a well trained and competent local research team will have the time and inclination to do this.

The education of such a well trained and strongly motivated body of applied scientists can be carried out only by people who are intimately acquainted with the frontiers of science in the fundamental fields and who are active research workers themselves. This is so because the fundamental science of today is the applied science of tomorrow and because the educational process in science is a long process. One or two examples should suffice to illustrate this point. It would have been proper, for instance, in the late 1930s, to include in the education of an engineer a rather thorough course in the then somewhat esoteric subject of nuclear physics, since this would have prepared these engineers to face, 10 or 15 years later, at the peak of their careers, the major developments in nuclear engineering connected with reactors. Experience has shown that it is very difficult for an already established scientist or technologist to interrupt his career to train himself in a fundamental branch of science which he missed when he was a student. Similarly, an electronic engineer receiving his training in the 1940s would have been well advised to learn a considerable amount of solid state physics, to prepare for the development of transistor technology in the 1950s.

That this point is not at all an academic one can be concretely illustrated. In most underdeveloped countries the few institutions of higher learning which exist are staffed with a large proportion of the very few local scientists or engineers who are available. Many of the more senior of these received their training in the thirties or forties at some foreign university, perhaps did some routine piece of work for a thesis, and then returned to the homeland, to be appointed to a post at a university or college. With no research being pursued there, with no competition from colleagues and in the general atmosphere in which time and change were of no great concern, they settled down to teaching the science of 1938 throughout the 1940s and 1950s, and now the 1960s. The concentration on the teaching of undergraduates accentuates this tendency to persist in the scientific ways of their own student days. By now, they have a vested interest in opposing any change in the curriculum, in the teaching methods or even in the administrative practices of the university, and since they are in senior positions, they act as a very strong barrier indeed to any modernisation of the institutions of higher learning. It is a very vivid example of what might be called scientific " featherbedding ", and it is strongly self-perpetuating inasmuch as the science students of today, who thus learn at best the sciences of yesterday, will get little of the excitement of being in the forefront of the sciences, will be badly prepared for a productive scientific career and hence will be mostly concerned with trying to secure for themselves the few comfortable old-fashioned academic positions which become available at the universities as the older generations disappear. To obtain these positions, they do not have to show great brilliance or present lists of research publications, but instead have to exhibit a willingness not to rock the boat and to fit in with the old-fashioned ways which these institutions have followed for the past decades. Almost all the universities on the Indian subcontinent suffer from this malady. The situation is somewhat different in the newer universities of Africa and South-East Asia but, for closely related reasons, very similar conditions obtain. With active research being pursued at these institutions, such a process of intellectual calcification could not possibly have come about and where it has not yet occurred it can still be avoided. If active and fundamental research is not carried on in these universities, this calcification will become or remain as firm as it already is in the older universities of the " third world ".

III

The second argument in favour of basic scientific research in underdeveloped countries is also linked with the applied work done there; it asserts that applied work can be carried out successfully only if the research workers have constant access to persons working in fundamental fields. There are several reasons for this. The training of a person working in fundamental research tends to be more thorough and broader than that of persons who committed themselves early in their careers to applied work. This is particularly true for scientific personnel in underdeveloped countries, where applied research is often defined in the narrowest possible sense of the word. For instance, a basic research worker usually has a better background in general research methods, in the understanding of the interconnection of various applied problems through their common scientific basis and in the general techniques applicable in various branches of a science. Thus, a person trained in fundamental research can often serve as an " ideas man " to his colleague trained in applied research, suggesting new approaches to a problem, or carrying over analogous methods or solutions from fields unfamiliar to the workers in applied research.

Such an interplay between fundamental and applied research is an essential ingredient in the advanced countries, so much so that perhaps we take it for granted that such opportunities exist in all countries. A very large fraction of research workers in the fundamental fields in the United States spend part of their time as consultants to applied projects for the government or private industry. They serve either as members of review committees, which survey progress made in a given applied field, and suggest new directions, new approaches and techniques, or, they might be called upon to help in connection with a specific applied problem which has run into difficulties and where a broader scientific base is helpful in getting it back on the right track. The results of such cooperation between applied and pure research workers are obviously very valuable and consultants of this kind are, therefore, in great demand and consultant's fees very high.

My point is simply that this assistance in carrying out applied research should also be available to the underdeveloped countries. In fact the need there is even greater since, once an applied team runs into difficulties, the closest colleagues who might come to the rescue may be thousands of miles and large amounts of foreign exchange away. Hence day-to-day cooperation is out of the question and even a single trip for the purpose of consultation might run into financial and political obstacles. The only remedy is to establish highly competent and active local research teams in the fundamental sciences. They can then be used as consultants in times of such " crises ".

IV

The third argument supporting pure research is not related to applied work but stands on its own merit. Granting that eventually all countries should have their own basic scientific life it will still take a very long time— perhaps several scientific generations—to establish a strong tradition of pure research in a country which has previously been without science. It should therefore be part of the planning of a new country to make, in addition to five-year economic plans, a 25-year scientific plan. This would allow for the training of young scientists today in the fundamental fields, who then, years from now, with considerable experience and some notable achievements behind them, will originate and direct research carried out by a new generation of local scientists. These, having been brought up in an indigenous research atmosphere, will be able to expand their own research efforts and in turn lead another scientific generation which will then be large and strong enough to carry out, on an extensive scale, research which will be competitive with work in the advanced countries.

There are several examples demonstrating that even in relatively advanced countries, the development of a scientific tradition of excellence is a slow process. It took even communist Russia, building on an already outstanding and established scientific tradition and on the scientific institutions of Czarist Russia, about 40 years to become a major scientific nation. As for the United States, it took a tremendous influx of European scientists, coupled with the scientifically extremely stimulating atmosphere of the war years, plus 20 years, to transform American physics from a minor and sporadic effort into world leadership. When we are faced with the meagre resources and considerable cultural handicaps of an underdeveloped country, such a process should be expected to take even longer. It is therefore of utmost importance to start it as soon as possible, even if on a modest scale.

V

The fourth point in connection with research in the less developed countries concerns the training of administrators. In the new states there is a tremendous need for competent and well-trained administrators to decide on the right policy for development and to carry out those decisions according to plan. This lack of local administrators has been cited as the largest single handicap of the entire foreign aid programme of the United States. Among these administrators, many have to make decisions which deal with scientific and technological matters such as health projects, new industrial processes, improved agricultural methods, development of power resources and many others. The need is very great, therefore, for persons sufficiently well versed in science and technology to be able to make technically sound, realistic and prompt decisions. The best training for such posts is research experience and a thorough education in the fundamental sciences. This contention is borne out by the example of the United States where the administrative heads of the large government and private research and development projects are increasingly scientists with research background in some fundamental branch of science. It should be kept in mind that especially in the case of a new country, where everything starts from a low level, where decisions affecting the country for several decades have to be made, and where furthermore the number of persons making the decisions is very small, the stakes are extremely high and a few mistakes might make the difference between success and failure. It should also be remembered that because of the shortage of scientific manpower these administrators in the young countries have no advisory committees to rely on for technical advice, as is the practice in the advanced countries, and they have no one to turn to for aid in making these decisions. It is of the highest importance, therefore, that these people get as broad an education in the various fundamental sciences as is possible and that they come to understand the ethos of science through their own experience in research. The respective Chairmen of the Atomic Energy Commissions of India and Pakistan, Dr. Bhabha and Dr. Usmani, might serve as individual examples of this contention.

VI

The fifth argument for fundamental research is a psychological one; achievements in the fundamental sciences would serve as a source of great encouragement and high morale in the newly developing countries. Justly or not, pure science is generally considered one of the most sublime proving grounds for the human mind and a country, which economically, socially and politically might still have to consider itself inferior to its Western counterparts, might take special pride in the outstanding achievement of one of its sons in the natural sciences. In a small country like Denmark, the late Niels Bohr is a national hero and one could encounter a taxi driver in Lahore, perhaps illiterate, who speaks with reverence of Professor

Abdus Salam, the most eminent Pakistani physicist. More importantly, such outstanding individuals also serve as models for young local scientists, who are trained under less than ideal conditions, or who, after having been abroad, have to face the agonising decision of whether to return to their backward homeland or to succumb to the many tempting offers of employment in the advanced countries. In making this decision and in gathering strength for hard work once they decide to return home, the example of outstanding local achievements in the fundamental sciences plays a crucial role. One cannot overemphasise the importance of high morale when talking about the development of backward countries. It may be the most important single factor in deciding between success and failure.

VII

In the foregoing arguments on behalf of fundamental research in underdeveloped countries, I have assumed at several points that the advanced education of scientists in a country should be carried out to a large extent on their home soil. The reasons for this are as follows.

Firstly, most of these developing countries have chronic shortages of foreign exchange and education abroad uses a lot of foreign currency. Reliance on fellowships can be only on a small scale; when hundreds of scientists have to be trained each year, it is impossible to find enough outside financial help. It is much more economical, even apart from its other benefits, to establish an institution of higher learning on home soil.

Secondly, Western institutions of higher learning are becoming more and more overtaxed by demands for admission of their own nationals, so that even if a candidate from an underdeveloped country is treated on an equal footing with his Western colleague (which, for a number of reasons, might not be the case), he has a good chance of being turned down. This situation will only be aggravated in the coming years with the rising demand for higher education in the West.

Thirdly, it is very difficult indeed to establish functioning research groups when people get their education in different university systems abroad, when the assimilation of new staff workers cannot be done at an early stage and when continuity of personnel is lacking. It has been a long-standing experience in all advanced countries that the best and cheapest hands in the gruelling work of scientific research is the postgraduate student who is responsible for a very large proportion of the many man-hours involved in scientific experiments. At the same time, the student himself learns, at an early stage, about the problems peculiar to the equipment he uses, about the sources of supply for spare parts, about the peculiarities of the supporting technicians, etc. All in all, a " school " thus develops around

170

a senior scientific person and this school can turn out to be extremely productive indeed. It should be the goal of an underdeveloped country to establish such schools on home soil in the various branches of the sciences and such schools cannot prosper without young students, especially postgraduate students.

Fourthly, living abroad during the advanced training and then returning home raises a number of problems of adjustment. For one thing, some trainees have been known to be unable to produce anything like their best in the Western countries because the problems they faced when trying to make an adjustment to the different economic, social and cultural practices were so overwhelming that their scientific studies were neglected. Others managed to adjust but in the process lost some of their sense of identity with their homeland and when the time came to return, they let themselves be lured away by the attractions of the research institutions of the countries in which they received their advanced training. In fact, the problem of how to encourage young Ph.D.s to return to their country of origin is one of the most serious in trying to develop science in a new state. The " mortality rate " is very high indeed and affects adversely the whole programme of advanced scientific training and research. Few governments would fail to grumble about spending foreign exchange to send their sons abroad only to be snatched away by the richer countries. These problems of adjustment and readjustment are really extraneous to science and hence should be eliminated if at all possible by establishing advanced institutions on home soil.

VIII

Let us now see in what way a technical assistance programme can be used to promote basic scientific activities in these countries. The three important ingredients in such activities are buildings, equipment and manpower.

The first of these, buildings, is the least difficult. Most of the young countries are well prepared, eager and willing to put up sumptuous edifices to house their research activities. There are several reasons for this. Firstly, buildings usually do not involve spending foreign exchange. To be sure, foreign architects are often engaged to make the designs but even some of these are willing to accept local currency and the architect's fee is a small fraction of the total cost anyway. The construction itself is done by a local firm, often not without faults or even corruption, but at least for the first few years the new building will serve as an imposing exhibit of the country's will to become " up to date ". And this is the second reason for the willingness to provide buildings: it is an obvious and eye-catching sign of " progress ", which can be used for impressing foreign visitors and for domestic propaganda. Only an expert, and only after a thorough investigation, can tell whether the work carried out in the building is high grade or not; the majority of the onlookers will judge the enterprise by its external manifestations, of which the building is the most evident one. The attractive buildings of universities in Mexico, Pakistan, or some of the West African countries may serve as examples.

The third reason why it is relatively easy to get buildings is that the financing of a building is a matter for a single decision, which might be made by the government in a burst of enthusiasm. Once built, it requires little further expenditure for maintenance. This is also one of the reasons why experimental apparatus, to be listed under the second ingredient, equipment, is relatively easy to get. Although this nearly always requires foreign exchange there are many advanced countries very willing and well prepared to donate it as part of foreign aid. This again has the advantage of looking impressive and in addition helping the industry of the donor country which produces the equipment. It also appears to have the advantage of involving only a single decision, although in reality this is not so. In fact, expert maintenance of such equipment is a major problem in the underdeveloped countries since the shortage of trained mechanics and technicians is even greater than the shortage of scientists and spare parts must also come from abroad. For this reason impressive equipment often stands idle for want of a very inexpensive replacement item or because nobody in the institution knows enough about the apparatus to find out what is wrong with it.

There are two immediate and easy remedies for this malady. The first would be a ready supply of spare parts in the country where the equipment is located. For instance, a storeroom could be set up which would carry all the spare parts needed for the equipment donated to all institutions in that country. Secondly, a system of roving Western mechanics could be organised until local ones have been trained. For example, one electronic technician would be assigned to Pakistan; he would constantly tour the universities, institutions, hospitals, etc. where equipment donated by the foreign government is located and could repair faulty equipment and give advice for the proper use of the apparatus. Considering the high cost of some of the electronic equipment, the cost of maintenance of this one technician would produce handsome returns in the increased efficiency of the apparatus.

The development of library facilities is another element to include in the category of equipment. This has at present a much less favoured position than experimental equipment, although it is hardly less important. Buying scientific books and periodicals requires foreign exchange and in fact, in large amounts, over a long period of time. Scientific libraries are generally not objects of national pride, being buried in one room of the upper floor of the building, and a layman generally has little basis for distinguishing a good library from a bad one. There have been several attempts made to facilitate the building up of libraries in these countries. Some books published in the United States appear in cheap "Asian editions ". There are some organisations, particularly the British Council, the Asia Foundation, and the United States Information Service, which occasionally donate subscriptions to periodicals, at least for a limited amount of time. But this is just a piecemeal solution on a small scale. The United States government should consider publishing all American scientific journals in cheap editions, which, aided by government subsidy, would be available to the underdeveloped countries either free or in exchange for local currency. At least one library in each city where considerable scientific activity is evident, should be well stocked with books, journals and research papers in temporary form, such as duplicated reports, so-called " preprints ", conference proceedings, etc. One of the greatest handicaps of a scientist working in an underdeveloped country is the feeling of isolation and this could be relieved by a prompt and ample flow of scientific information. (Other countries with technical assistance programmes could also help by supplying scientific information along the same lines.)

The most serious problems of science in an underdeveloped country are, however, those connected with the supply and maintenance of manpower of high quality. I have already discussed some of the problems in connection with giving advanced scientific education to promising students and touched on the difficulty of luring back trained personnel from abroad. Once promising young scientists return to their homeland, they are faced with further problems, such as delays in obtaining experimental equipment and problems in maintaining it, inadequate library and other information facilities, lack of contact with their colleagues in the more advanced countries, shortage of stimulating local colleagues working in the same field with whom to discuss problems, low salaries, too many routine duties, etc. These are problems which arise from national poverty, small numbers of trained personnel, faulty traditions and uncongenial administrative practices.

IX

Clearly, many of these problems have to be tackled one by one and in the context of the local conditions. In addition, however, there is also a general remedy, which might be supplied within technical assistance programmes, and this I wish to discuss in greater detail. Life in the fundamental branches of science would be greatly stimulated if active research workers from advanced countries would be willing to visit research institutes in the developing countries for a stay of, say, one year, to cooperate with the local workers in their research and organisational activities.

What would such scientists bring to an underdeveloped institution that is unique and valuable?

First, they would bring up-to-date scientific methods, information on the latest developments and up-to-date ideas. The isolation of institutes in the young countries could be greatly reduced by the wealth of information infused by the presence of persons who have lived in the most stimulating environment of one of the great Western centres of research. Such visitors' programmes are essential even for those Western research establishments which are slightly outside the mainstream of activities. They are thus even more vital in those institutions whose only contact with the latest developments of contemporary science would otherwise be through the printed page.

Secondly, one of the striking characteristics of the scientific life of the underdeveloped countries is the fact that the personnel which counts, as far as modern research is concerned, consists mostly of quite young persons. This is to be expected because higher education in most of these countries, particularly in the natural sciences, is very recent. The consequence of this is, however, that there is a crying need for originators of research, for "ideas men", for those who can stimulate and guide the young Ph.D.s until they have acquired enough experience themselves to carry out independent research. Only a very small fraction of new Ph.D.s are sufficiently brilliant and original to be able, immediately after receiving their degree, to originate research problems or lead a research team. In advanced countries this does not pose a problem, since these young people in effect get a long post-Ph.D. training as junior members of some research team or academic staff. In an underdeveloped country, where at present such teams and staff do not exist, a research worker from a leading university or research institute in a scientifically advanced country with some experience in generating research problems and in leading research, could do wonders in making effective use of local talent which otherwise might lapse into aimless stagnation.

Thirdly, a Western research worker, even if he has no special interest or experience in scientific organisation, is likely to have acquired unwittingly the knowledge of how an advanced research institution functions. Such tacit knowledge is badly needed in the underdeveloped countries. To be sure, some of the local scientists themselves have received their advanced scientific education at a Western research institution. Their adjustment problems, however, as well as their position as foreign guests, usually prevent them from learning much about the organisation of the institute in which they are being trained. Their task is to do well in their professional studies and they consider their mission a success if this specific aim is fulfilled. There is, hence, a great need in the young research establishments in the underdeveloped countries for advice on matters of curriculum, library practices, training, organisation of seminars, store and workshop rules, etc. A particularly important aspect of this in which a Western scientist can be of great help is the attraction of short-term visitors, persons who happen to pass through the area and are willing to visit for a few days, conduct a seminar or two, and talk to staff members about their problems. Travel for scientists in the Western countries is recognised as imperative and an increasing number of scientists are going beyond the California-Moscow circuit and visiting countries in Africa, South Asia and Japan. Usually only a very small additional amount of money is needed, in local currency, to persuade the visitor to make a small detour and visit a given institution. But the contact must be made by somebody who knows (through the scientific grapevine) that the visitor is coming and who knows him well enough to invite him. A Western scientist, particularly if he has very extensive personal contacts, is invaluable in this respect.

In addition to these rather tangible benefits, the Western visitor represents, much beyond his role as a scientist, a note of encouragement, a stimulus to morale, a symbol of recognition, which is of extreme importance. In luring back young trainees from abroad, the assurance that visitors from the advanced countries will be joining them might be a decisive factor. A Western visitor can also enlarge the research group which otherwise would be too small to function properly. Finally, the visiting scientist from an eminent institution will be a symbol that the institution in question has " arrived ", that it is, in some sense, on an equal footing with its foreign counterparts as a member of the world scientific community, that it has established a strong link with the international brotherhood of scientists. The feeling of elation over this development and the high morale generated by it will compensate for many of the physical shortcomings of the research establishment and will give strength to the local scientists for the hard work facing them.

X

Thus far, this form of technical assistance, namely, the sending of research scientists for prolonged sojourns to an underdeveloped country has hardly been explored at all. The problem as such has not received adequate recognition. But even when the problem is recognised, the task will still remain more difficult to organise than the sending of equipment or the building of a dam, since it involves a large group of scientists and technologists, who will have to be convinced that the programme is valuable and they will also have to be convinced that they can assist in it without undue personal sacrifice. It requires convincing the research scientist in the United States and in other countries with technical assistance programmes, working at a national laboratory, university, or private firm, that he should transfer his own work to a remote country, at least for a year, where he might have to spend some of his time getting adjusted, or serving as an adviser to the local scientists, and where consequently his efficiency as a research worker might suffer. With some advanced planning, however, and by choosing the appropriate time for the visit in relation to the career of the visitors, such a temporary adventure might be not only not detrimental but in fact beneficial to the visitor, even from the purely scientific point of view. His year abroad might be spent catching up with broader developments in his field, with devising plans for future work, with writing a book, or just with getting a new perspective on the problems with which he has been confronted in his daily routine at home.

Channels of financial help for such an undertaking are at the present very meagre, at least compared to those available for educational projects (Fulbright, Smith-Mundt, etc.) or for technical cooperation (AID, Point Four, Colombo Plan). It would be a great help if special fellowships could be set up for this purpose, perhaps by the National Science Foundation in the United States and corresponding bodies in other countries, so that an interested scientist would not have to take on himself the responsibility for seeking financial support. Two hundred such fellowships annually, costing say, $25,000 each, would only amount to $5,000,000 a year, or a fraction of 1 per cent. of what is now spent on technical aid. Yet, 200 scientists abroad distributed over 50 underdeveloped countries which are ready to benefit from the programme outlined, would make a tremendous impact indeed on fundamental research all over the world and would in the long run pay a very great return in the form of the scientific results obtained from the institutions and the countries to which they go.

SOME PRACTICAL SUGGESTIONS FOR THE IMPROVEMENT OF SCIENCE IN DEVELOPING COUNTRIES

Michael J. Moravcsik

In a previous article,[1] I discussed some of the reasons for my belief in the importance of a newly developing country creating a firm foundation not only in applied but also in basic research in the natural sciences. Following that general discussion, I would now like to turn to the listing of some specific and concrete steps that can be taken to achieve the general goal.

It might be well to begin my discussion of a controversial problem with a note of caution. My own personal experience with the problems described in this article has been gained mainly (although not exclusively) during the academic year 1962–63, which I spent in Pakistan as a temporary employee of the International Atomic Energy Agency, assigned to the Pakistani Atomic Energy Commission's Atomic Energy Centre in Lahore. It is possible that this experience does not allow an immediate generalisation to other developing countries with completely different historical, cultural and political backgrounds, such as Nigeria or Uruguay, and thus my conclusions might be valid mainly for the Indian subcontinent.

When a Western " expert " visits a university in a developing country, he often finds himself requested to review the local syllabus or curriculum and to suggest changes. He also often hears complaints about the absence of textbooks and other instructional aids. Indeed, there is much to be done to improve the situation in these respects. But he is also likely to come to the conclusion that the main trouble with science education at these universities lies in a different domain which is more difficult to convey to the local staff and even more difficult to remedy. The problem is not with *what* is being taught but with *how* it is taught. There is an overwhelming tendency on the part of the staff and the students to consider science as a collection of facts to be memorised, an abstract discipline consisting of general laws of nature, the faultless recitation of which makes a well-qualified scientist. That the best way to absorb science is to work out problems is very little appreciated and hardly ever practised. I have encountered undoubtedly very bright young men in Pakistan who represented the top of their class at the best Pakistani universities. and who

[1] " Technical Assistance and Fundamental Research in Underdeveloped Countries ", *Minerva*, II, 2 (Winter, 1964), pp. 197–209.
This article from *Minerva*. IV, 3 (Spring, 1966), pp. 381–390.

could rattle off the second law of thermodynamics much better than I can, but who, when confronted with the question of what happens to an insulated room when a working refrigerator with an open door is placed in it, not only did not know the answer, but failed to realise that this " silly " question had anything to do with thermodynamics.

What can be done to foster the transition from learning by rote to the perception and selection of problems among those who have concluded an undergraduate course in physics and who would like to go on to postgraduate work abroad? The Atomic Energy Centre where I was stationed in Lahore has a training programme for its young employees (most of whom have received an M.Sc. degree at one of the Pakistani universities) prior to sending them abroad for further training. This programme now includes three courses, in mechanics, electrodynamics and modern physics, respectively, based on American textbooks I recommended; the books and the courses lay heavy emphasis on problem-solving. Courses like these might make the transition to the style of work of a Western graduate school less painful. Not less important would be the experience which local teachers would gain in teaching in a more problem-oriented manner.

Postgraduate Education

Although it is very desirable in the long run for an underdeveloped country to be able to train its scientists at home on the postgraduate level too, this is likely to come only at a relatively advanced stage. In the meantime, such training must be obtained at Western graduate schools. This raises many problems.

The very first one has to do with obtaining admission and financial support at a Western graduate school. The pressure for places has been increasing every year and the better schools sometimes have 10 times as many applicants as they are able to admit. In such a situation an applicant from an underdeveloped country has many handicaps from the very beginning. Quite often the record of his predecessors in that department is not too distinguished. Furthermore, his application is supported by academic records which tell very little and by glowing letters of recommendation from completely unknown teachers, who often consider it a feather in the cap of their own prestige if another student is admitted to a Western institution. Since such an applicant represents so many question marks, he is most likely to be turned down by the graduate school which cannot afford to take chances with its scarce places.

There are several possible remedies for this problem. An admittedly stop-gap measure is the one I have tried with trainees of the Pakistan Atomic Energy Commission. I have made arrangements with about a dozen good American graduate physics departments, under which I help

178

to assure them that the Pakistani students I recommend to them are in fact prepared for Western postgraduate physics education. In return the schools are willing to consider the applications of these students in the light of this additional information, although, of course, this does not mean automatic admission and financial support. My own recommendation, in turn, is arrived at in the following fashion. First, there are some Pakistani colleagues whose judgement I have come to rely on and who can compare for me the new applicants with previous applicants, some of whom might already be successfully engaged in postgraduate work in the United States. Second, I have prepared a written examination (consisting mainly of problems), based on the A.E.C. training programme I mentioned in connection with undergraduate education. This examination is then given by a colleague in Pakistan and is forwarded to me for review.

This method is admittedly awkward, piecemeal, and cannot be expected to work forever, as my personal contact with the Atomic Energy Centre becomes attenuated. One would think that some universal written examination could be worked out, similar to the Graduate Record Examination, which could serve as a fair indication of the candidate's ability and preparedness. Although several organisations are working on the construction of such examinations, I doubt very much that they will be successful. Because of the great discrepancy between the educational methods of the different parts of the world and because of the specific shortcomings I discussed above in connection with undergraduate education, I believe that such an examination would indicate very little.

My suggestion for a more permanent and more efficient solution of the problem of selecting students for postgraduate education in the West is the establishment of *ad hoc* interviewing committees which would visit the various universities and interview, for an hour or so, each candidate who wishes to apply to a Western graduate school. A committee of three physicists, travelling for one month, could probably take care of all applicants for physics postgraduate education in all the Asian countries, at a total cost of less than $15,000. It has been my experience that an hour-long interview offers an excellent opportunity to make a fairly reliable judgement of the applicant's accomplishments, capacities and weaknesses. This assessment, should he be admitted to a foreign graduate school, would be valuable information for his supervisor there. The information obtained by this committee would, of course, be made available to any university which might be interested in it.

During the past two years I have been trying to persuade a number of American foundations and government agencies to support this scheme and, while everybody seems to be eager to see it carried out, actual financial support has not yet materialised.

The next question that arises in connection with graduate education is related to the branch of science in which a student should be trained. In order to be able to offer the student, at the completion of his education, a position at home in the proper field, he must be trained in a field which is in fact being worked on at home. At the same time, practice has shown that a simple order prescribing the field of study for a student never works, and perhaps rightly so. The rather obvious solution of this problem is to place the student in a Western university where the field of science which his home university or laboratory has in mind dominates the department and where the student is likely of his own volition to choose that field. This arrangement should be followed even if a particular university is perhaps not as excellent in an overall sense as another in which the particular field is less strongly represented. Since the worship of " great names " among universities is even more pronounced in developing countries than in advanced ones, this procedure requires a certain amount of self-restraint. It is, however, indispensable because it is extremely difficult to change the field of interest of a student once he obtains a Ph.D. in that field.

The Effectiveness of Postgraduate Study

Now let us assume that the student has been able to find an appropriate university, obtained admission there and received financial support. What can be done at this stage to make his transition to his new environment easier and more effective?

First, it must be realised that in most cases such a student will need a lot of individual attention from his adviser and that this attention and cooperation must often be initiated by the adviser since the student is too shy or confused to do so. One of the most important functions of the adviser is to ensure that the student does not take too heavy a load of postgraduate lectures and seminars if his undergraduate background is not absolutely sound. More often than not, such a student will in fact need one or two terms to take some advanced undergraduate subjects to catch up with his Western contemporaries. He will consider the taking of such courses not only a personal affront (being used to being the much-admired top student of his class) but might even consider it a tragedy since it might delay the completion of his doctorate by a year. The emphasis on the formal aspects of education in the developing countries manifests itself not only through the formalistic attitude toward science teaching and learning but also in the excessive concern with degrees and titles. Even if the student's financial assistance is assured throughout his course of postgraduate study, he will want to obtain his degree in the shortest possible time and is sometimes willing to disregard all other considerations, including the quality of his education, to achieve this. Nevertheless, the adviser should make certain that the student has a very solid working knowledge of the basic subjects before he permits him to enter upon the advanced phases. By now, many of the larger physics departments have " experts " on the staff who have dealt with many foreign graduate students, but even then, dealing with some of these cases might require an inordinate amount of time which the adviser, himself a working research scientist, is not willing to give. A judicious allocation of such foreign graduate students among those in the department who realise the problems involved might in some cases " save " students who, if left without special supervision, would fall by the wayside.

Once the student has completed his work towards an advanced degree and received his diploma, he is confronted with the question of whether to return home or not. This is one of the thorniest problems, which will be solved only when postgraduate education on a high standard is established in the developing country itself. In the meantime, there are various measures that should be taken to encourage the young Ph.D. to return.[2]

The most important requirement is that a young scientist must have a good opportunity for research at a respectable salary in his home country. It cannot be expected, and in fact it is not even necessary, however, that these facilities or the salary match those that exist in some places in the West. For one thing, most such scientists would not be at the leading institutions in the West even if they did not return home, and to match the facilities of second- or third-class institutions in the West is not an insurmountable problem. As far as salaries are concerned, what seems to matter is not the absolute standard of living but the standard with respect to the rest of the population. Thus a salary of Rs. 1,000 in Pakistan is a very respectable one, although its dollar equivalent in the United States would be considered quite poor. I believe it is generally true that a young man, who lived in his native country up to the age of 21 or so and then has spent four years in getting a graduate degree in a Western country, prefers, other things being equal, to return to live in his own country. By " other

[2] One of these, among others, is the education on this matter of American public opinion which at the present sees the issue from a very one-sided point of view. It is not entirely uncommon for a well-qualified young scientist from a developing country, upon receiving his degree in the United States, to seek the help of newspapers and other information media to obtain permission to continue to stay in the country. The case is usually described in terms of a brilliant young scientist whose skills are greatly needed by the United States, whose training would be completely wasted in his homeland, and who might even be persecuted upon his return to his native country which is described as ruled by a non-democratic government. It is claimed that the United States Immigration and Naturalization Service, because of its bureaucratic outlook and cumbersome procedures, forces the poor student to leave the United States. Sometimes, for local political reasons, a congressman enters the case and the deportation is delayed, often long enough for the student to marry a young American woman, in which case the matter is settled. I believe that if the other side of this coin were explained to the American public in greater detail such incidents would be less frequent.

things being equal" I simply mean adequate research facilities and adequate financial support in the above sense.

Often the student from the underdeveloped country who wishes to stay abroad is not " brilliant " enough to be really in great demand in an advanced country, but his competence could be very valuable in his own country where persons with his kind, quality and amount of training are few and where his own prospects are, therefore, greater. One of the really exciting aspects of life in a developing country is that there a single person with good training and much energy and determination (and, of course, some support from the local leadership) can have a tremendous impact on the country as a whole. In the West, where well-trained, highly intelligent people appear in great numbers, most of them have to be contented with making themselves felt only on a rather microscopic scale. It is therefore in the best interest of everybody (including the scientist in question) if he is given an opportunity somewhere where his contributions can be maximised.

It has sometimes been a practice of governments of developing countries to require a bond from the student just about to be sent abroad, guaranteeing his return after he has received his degree. I do not think this is a good practice. For one thing, it simply does not work, since the bond is small enough (*e.g.*, $2,000) so that if the student did stay in the West, he could pay it off within a short time without prohibitive sacrifices. Secondly, such a bond establishes a rather strange relationship between the student and his home country, in which the former's obligations are expressed in purely financial terms. An alternative method would be for the Western scientist who arranges the admission or the head of the department to which the student is admitted to ask the student for his word of honour that he will return to his country for at least two years after getting his degree. This obligation could still be changed if really unusual circumstances arose but it would make the whole matter a part of the personal relationship between the student and a Western scientist, something that would undoubtedly appeal to many students.

Overcoming Intellectual Isolation

Some of the problems which face working research scientists in developing countries were discussed in my first article. The most important handicap of a scientist in a developing country is his relative isolation from contact with other scientists. The inferiority of physical facilities, although an important factor, is definitely less crucial and also much easier to remedy. I would like, therefore, to suggest five ways in which the isolation of these scientists can be relieved.

The first of these might be to grant rather frequent sabbatical years at

Western research centres. It is not unreasonable to expect that, in order to function near his capacity, a scientist in a developing country should spend one out of every three years in the West. The problem here is twofold, first to find positions for these scientists and second, in view of the market for scientific skills in Western countries, to dissuade institutions in those countries from trying to lure these scientists away from their own countries on a permanent basis. With respect to the former, one of the difficulties in finding a temporary position for such scientists is the same that arises with gaining admission for students at graduate schools: the lack of reliable references. As in the case of the students, the best type of reference is the one established by personal contact. This in turn can be brought about by increased travel of Western scientists to research establishments in the developing countries and by increased participation of scientists from the developing countries at international conferences, summer schools and other research meetings. Many universities in the United States have by now developed a considerable tradition and interest in offering temporary appointments to scientists from developing countries, but much more needs to be done, possibly with financial assistance from the United States Government. A significant step in the right direction has been the recent establishment of the International Centre of Theoretical Physics in Trieste, Italy, which is heavily slanted towards the temporary accommodation of scientists from developing countries.[3]

A second specific step could be the establishment of foreign-financed regional research centres with a significant number of foreign scientists on the staff. Although participation in such international centres might be hampered by local political considerations, the equitable distribution and impartial direction of such centres could be arranged under UN or other auspices. For example, two such centres in physics (one in Pakistan, and one in India, one specialising in solid state physics, the other in atomic and nuclear physics) would benefit not only those two countries but several others in that part of the world who, at present, are too undeveloped to do much on their own in scientific research. Such a centre would have all the advantages of being *in situ*, thus catalysing the scientific life of the region considerably beyond the primary effect it might have on the few scientists on its staff.

The third improvement involves the visits of foreign scientists to the research establishments in developing countries. (Since I dealt with this point at some length in my first article, I will not discuss it here in more detail.)

The fourth field where help can be provided has to do with written contact with the rest of the scientific world. The matter of books and

[3] *Cf. Minerva,* III, 4 (Summer, 1965), pp. 533–536.

journals was discussed in my first article. But in many branches of science (including the one I happen to be working in, namely elementary particle physics) books and journals are more and more relegated to being a depository of completed research for the purposes of later review or for access by later generations. The current research results, the " breakthrough ", the " hot arguments " are propagated by conferences, personal letters and, above all, " preprints ", which are rapidly duplicated copies of research papers, just completed, which might appear in journals six or 10 months hence. The distribution of these " preprints " is often done rather haphazardly and there is a tendency to flood well-established persons at leading research centres while omitting altogether little known scientists in less famous scientific establishments (who often need the preprint most). A new programme, soon to be put in operation, will centralise the duplication and distribution of such preprints, at least in high energy theoretical physics, and will send them to any group of scientists working in this field anywhere in the world, free of charge. Such a programme in other branches of science would be of great value in stimulating research in the developing countries. It is difficult for anyone not engaged in research in the pioneer fields of science to appreciate the psychological uplift and increase in research effectiveness brought about by the knowledge that vital information reaches the research worker simultaneously with those in the advanced countries.

Finally, cooperative research between Western and other scientists might also stimulate the scientific life in underdeveloped countries. A young scientist, returning home after recently obtaining his degree, often lacks the perspective to choose interesting problems to work on. But even for more experienced scientists, a congenial colleague interested in the same field is often lacking. In such situations, cooperation with a Western scientist on a certain research project might be very useful. Communication is possible by mail, although, of course, this is not at all as effective as personal contact. Such cooperation is clearly more feasible in theoretical research than in experimental work. It requires a certain amount of time on the part of the foreign half of the team, since writing out everything on paper is often time-consuming, but the rewards are often gratifying.

As I mentioned at the beginning of this section, the problem of physical equipment is also an important one. I would simply like to repeat here the suggestion made in my earlier article that a very significant improvement could be made in the efficient use of already existing equipment in the developing countries by establishing a programme of roving Western technicians, well equipped with spare parts, who would make sure that a $10,000 piece of apparatus does not lie idle for six months for the want of a 10 cent part and the know-how for its replacement.

Conclusion

Some of the above suggestions, if implemented, could contribute substantially to the stimulation of scientific life in the developing countries. In conclusion, however, two general points should be emphasised.

The first is that all the measures recommended above require patience and perseverance. It seems to be true that the less developed a country is, the more conservative its people are and the less amenable they are to changes in their lives and habits even if, from an "objective" point of view, such change is clearly "to their advantage". The most important barrier to cross therefore when trying to contribute to any aspect of the development of an emerging country is to conquer the apathy, the inertia and the lack of urgency that generally prevails. Usually a few talented, visionary and energetic local figures will work hard at bringing about such changes but the results will be sometimes uncertain and almost always slow to emerge. That such slowness is in the nature of things is important to realise in order to avoid the disillusion and demoralisation that I have often seen among Westerners working in the developing countries and even sometimes among the more enlightened local leaders. But coercive methods can do nothing, and least of all in scientific research. A middle road can be found which accepts and seeks slow but steady change, fast enough to make considerable progress in the long run but not so fast as to make whole generations of people permanently miserable and insecure.

Finally, it must be stressed that the most important general factor in the success or failure of the scientific life of a developing country is the morale of its individual scientists. I have seen over and over again scientists from those countries, who worked hard and produced interesting results while staying in a Western country, falter and fail when returning home to research conditions which were by no means worse than those they had in the West. Partly, it might be the influence of the environment, the sight of too many people sitting around in the streets, doing nothing. Partly, it might be a secret conviction that it is just impossible to do research in an underdeveloped country. In part, it might also be due to their private life in a society whose standards, mores and values are different from a modern industrial society. In any case, it is primarily a psychological problem and, as such, might often be contrary to the norms of a "rational" analysis. At the same time, small and objectively speaking insignificant factors can often cause great improvements in the morale of a scientist in a developing country. An invitation to a conference, a Western visitor for a week, a joint paper with a Western scientist, being placed on the list of those who receive preprints, or even repeated reference in the literature to work done by him all contribute towards dispelling the feeling that he is

185

excluded from the community of scientists, that he is cast out into the darkness where the handicaps are insurmountable. These are all steps that are easy to take but are often not thought of because their significance is not appreciated. And yet, in the last analysis, it is the enthusiasm and high morale of the leading individual scientists that will determine the rate of scientific progress in the emerging countries and hence will decide whether the gap between the advanced and emerging countries will continue to grow or whether it will begin to close.

THE GROWTH OF SCIENCE IN SOCIETY

Michael Polanyi

The Warrant of Scientific Judgement

The current situation in the philosophy of science is a strange one. The movement of logical positivism, which aimed at a strict definition of validity and meaning, reached the height of its claims and prestige about 20 years ago. Since then it has become clearer year by year that this aim was unattainable. And since (to my knowledge) no alternative has been offered to the desired strict criteria of scientific truth, we have no accepted theory of scientific knowledge today.

Most writers on science now accept the validity of science as unquestionable and neither in need of philosophic justification nor capable of justification. You will rarely find this spelled out, but it is revealed by current practice. Take Ernest Nagel's widely accepted account of science.[1] He writes that we do not know whether the premises assumed in the explanation of the sciences are true; and that were the requirement that these premises must be known to be true adopted, most of the widely accepted explanations in current science would have to be rejected as unsatisfactory. In effect, Nagel implies that we must save our belief in the truth of scientific explanations by refraining from asking what they are based upon. Scientific truth is defined, then, as that which scientists affirm and believe to be true.

Yet this lack of philosophic justification has not damaged the public authority of science, but rather increased it. Modern philosophers have excused this unaccountable belief in science, by declaring that the claims of science are only tentative and ever open to refutation by adverse evidence. And this has added to the authority of science. It was taken to show that, while scientific knowledge was supremely reliable, scientists were at the same time supremely open-minded, setting thereby an example of incomparable modesty and tolerance.

The Velikovsky Affair

Still, there have been some occasions in this century when the very foundations of science have been challenged. Sporadic attacks came in the form of laws against teaching the theory of evolution and in a papal

---

[1] Nagel, Ernest, *The Structure of Science* (New York: Harcourt, Brace and World, 1961), p. 43.
This article from *Minerva*, V, 4 (Summer, 1967), pp. 533–545.

encyclical warning against the evolutionary theory. An attack on a broad front was made in the Soviet Union. But a more personal challenge to the foundations of science came from Dr. Velikovsky's highly unorthodox book, *Worlds in Collision*, published about 15 years ago.[2] This book was emphatically rejected by scientists and yet it had a wide response among the lay public, who much resented the summary dismissal of the book by the experts. Moreover, the conflict widened, when some three years ago, new evidence turned up supporting Velikovsky's theory and this was again bluntly rejected by scientists. This seemed utterly unjustified to many people and a group of social scientists took up the matter. In *The American Behavioral Scientist*,[3] under the leadership of its editor, Dr. Alfred de Grazia (professor of social theory in government in New York University), they launched a systematic attack on the whole procedure by which contributions to science are tested, accepted or rejected.

A few words may cover Velikovsky's theory for my present purpose. The theory is based on the acceptance of evidence from the Old Testament, the Hindu Vedas, and Graeco-Roman mythology about the occurrence of catastrophic events in the earth's history from the fifteenth to the seventh century B.C. It interprets these disasters and upheavals as due to the repeated passage of the earth through the tail of a comet. This comet, Velikovsky claims, subsequently collided with Mars and by losing its tail transformed its head into the planet Venus. Further terrestrial upheavals ensued when in the year 687 B.C. Mars nearly collided with the earth; on one occasion the earth turned completely over, so that the sun rose in the west and set in the east. To account for these events Velikovsky supplements Newtonian gravitation by the assumption of powerful electrical and magnetic fields acting between planets.

As I said before, these ideas were rejected by astronomers, but they appealed to a wide circle of laymen: the book actually became a best-seller. And so bitter was the reaction of astronomers and other scientists to this, and such was the pressure they exercised, that Macmillan, who had published Velikovsky's book, felt compelled to give up their rights in it. They passed them on to Doubleday who felt less vulnerable to the hostility of scientific opinion.

When eventually, in September 1963, *The American Behavioral Scientist* published its protest against the treatment of Velikovsky, the editor of the journal, Professor de Grazia stated the aim of the inquiry

---

[2] Velikovsky, Immanuel, *Worlds in Collision* (New York: Macmillan, 1950).
[3] *The American Behavioral Scientist*, VII, 1 (September, 1963), contains a foreword on " The Politics of Science and Dr. Velikovsky " and articles by Ralph E. Juergens on " Minds in Chaos: A Recital of the Velikovsky Story ", Livio C. Stecchini on " The Inconstant Heavens: Velikovsky in Relation to Some Past Cosmic Perplexities ", and Alfred de Grazia on " The Scientific Reception System and Dr. Velikovsky ".

undertaken by him as follows: " The central problems are clear: Who determines scientific truth? What is their warrant? " and he added that " some judgement must be passed upon the behaviour of scientists and, if adverse, some remedies must be proposed." [4] Referring to the three articles in this issue of his journal de Grazia went on to say:

> If the judgement of the authors is correct, the scientific establishment is gravely inadequate to its professed aims, commits injustices as a matter of course, and is badly in need of research and reform.[5]

Now, if we accepted the modern critique of science, which leaves us today no other ultimate criterion of a scientific teaching than that scientists accept it as valid, the answer to de Grazia's first question, " Who determines scientific truth? ", would be simply: " The scientists." And to the second question, " What is their warrant? ", we would answer that the decisions of scientists cannot be accounted for.

But this is unacceptable. We must be able to say whether the reception of Velikovsky's ideas by scientists was fair and, if not, what went wrong? To find out this, Professor de Grazia sets out some points of a rational procedure for testing a proposed contribution to science. A contribution must not be rejected unread; its author may claim that it be tested and publicly discussed with him; if the contribution suggests radical innovations, these should be welcomed; if its ideas were at first rejected, the author should have a chance to return with additional proof and his claims should be once more cordially examined, and no authority should prevail against experimental evidence.

Professor de Grazia shows that all these rules were broken in Velikovsky's treatment by scientists. His work was condemned as utter nonsense by distinguished astronomers who frankly said that they had not read his book. He asked to be admitted to a public discussion of his views and this was refused. He had concluded that the surface of Venus was hot and its atmosphere heavy with hydrocarbons and asked the Harvard Observatory to test this prediction: this was refused. This happened in 1956. In February 1963, the American space explorer, Mariner II, confirmed Velikovsky's predictions about Venus: its surface

---

[4] *Ibid.*, p. 3.
[5] *Ibid.*, p. 3. The present paper was completed before the publication of de Grazia, Alfred (ed.), *The Velikovsky Affair. The Warfare of Science and Scientism* (New York: University Books Inc., 1966), which expands the contributions to the *American Behavioral Scientist*. The article by Professor de Grazia, to which I have referred in my essay, is revised and the passages on " the central problems " are now to be found in de Grazia's introduction to the book, while the other passage I quoted verbatim is absent. Though I can find no substantial shift from the position stated in *The American Behavioral Scientist* I would say I have come to appreciate better the merit of Professor de Grazia in raising the question concerning the grounds on which contributions to science are accepted by scientists.

temperature was 800° F. and its clouds appeared replete with hydro-carbons,[6] but this confirmation of the theory did not succeed in causing its discussion to be reopened by scientists; it was rated as a curious coincidence. Authority prevailed against facts.

It is understandable that Professor de Grazia was disappointed by the failure of scientists to live up to their professions to give a ready hearing to any new ideas and to submit humbly to the test of any evidence contra-dicting their current views. No wonder, perhaps, that he then went on to suggest that the acceptance of a new contribution by science may not depend on the evidence of its truth, but takes place either at random, or in the service of ruling powers, or in response to economic or political interests, or simply as dictated by accepted dogma.

De Grazia was right in contrasting the principles which scientists profess to follow in treating a novel contribution to science with the way they treated Velikovsky's ideas; but these principles must not be applied literally. They should be qualified by their tacit assumptions, and, once done, this resolves not only the anomalies of the Velikovsky case, but also the dilemma that the current philosophic critique of science leaves the validity of science unexplained. For both these anomalies are due to the fact that the demand for strict criteria of scientific truth causes us to overlook the tacit principles on which science is actually founded. Let me show these tacit operations.

The Tacit Component in Scientific Judgement

A vital judgement practised in science is the assessment of *plausibility*. Only plausible ideas are taken up, discussed and tested by scientists. Such a decision may later be proved right, but at the time that it is made, the assessment of plausibility is based on a broad exercise of intuition guided by many subtle indications, and *thus it is altogether undemonstrable. It is tacit.*

To show what I mean, let me recall an example of a claim lacking plausibility to the point of being absurd: I found it many years ago in a letter published in *Nature*. The author of this letter had observed that the average gestation period of different animals ranging from rabbits to cows was an integer multiple of the number π. The evidence he produced was ample, the agreement good. Yet the acceptance of this contribution by the journal was only meant as a joke. No amount of evidence could convince a modern biologist that gestation periods are equal to integer multiples of π. Our conception of the nature of things tells us that such a relationship is absurd.

---

[6] See Juergens, Ralph E., " The Evidence from Mariner II ", *The American Behavioral Scientist*, VII, 1 (September, 1963), p. 7.

A more technical example from physics can be found in a paper by Lord Rayleigh published in the *Proceedings of the Royal Society* in 1947.[7] It described some fairly simple experiments which proved in the author's opinion that a hydrogen atom impinging on a metal wire could transmit to it energies ranging to a hundred electron-volts. Such an observation, if correct, would be far more revolutionary than the discovery of atomic fission by Otto Hahn in 1939. Yet when this paper appeared—and I asked for various opinions about it—the physicists only shrugged their shoulders. They could not find fault with the experiment, yet they not only did not believe its results, but they did not even think it worth-while to consider what was wrong with it, let alone to check it; they just ignored it. About 10 years later, some experiments were brought to my notice which accidentally offered an explanation of Lord Rayleigh's findings. His results were apparently due to hidden factors of no great interest, which he could hardly have identified at the time; he should have ignored his observation, as his colleagues quite rightly did.

I have also had occasion to describe how the possible reality of a certain type of observation which had long been denied, was for a time accepted, then again rejected, only to be soon accepted again and presently rejected once more—these two consecutive alternations of acceptance and rejection taking place within 25 years. The observations in question were the apparent transformations of elements. Ever since the immutability of elements had been accepted, such observations had been cast aside as " dirt effects ". After Rutherford and Soddy established the fact of radio-active disintegration, such effects were taken seriously and accepted for publication—but disappeared from the journals again when it was recognised that radioactivity occurs only in a very few, comparatively rare elements. But new reports of apparently well-authenticated cases appeared in journals once more in response to Rutherford's discovery of the artificial disintegration of elements; and these presently vanished again, as the nature of such disintegration became clear and showed that it could not happen in a chemical laboratory. Apparent chemical transformations of elements have no doubt continued to turn up, and have been unhesitatingly ignored as they had been for so long until the advent of radioactivity.

The Tacit Component and the Dangers of Misjudgement

Suppose then that Velikovsky's claims were as implausible as the parallelism between periods of gestation and the number π; or as implausible as Lord Rayleigh's results published in the *Proceedings of the*

---

[7] Rayleigh, Lord, " The Surprising Amount of Energy which can be collected from Gases after the Electric Discharge has passed ", *Proceedings of the Royal Society*, Vol. 189 (May, 1947), pp. 296-299.

Royal Society; or, again, as implausible as the chemical transformation of elements now appears to be—or that they were to appear even more absurd than these claims—then it would certainly correspond to the current custom of science to reject them at a glance unread, and to refuse to discuss them publicly with the author. Indeed, to drop one's work in order to test some of Velikovsky's claims, as requested by him, would appear a culpable waste of time, expense, and effort.

But how about the predictions of Velikovsky which came true? Should these not have caused his book to be reconsidered? No, a theory rejected as absurd will not always be made plausible by the confirmation of some of its predictions. The fate of Eddington's cosmic theories may illustrate this. In 1946, I put on record an anxious remark by a distinguished mathematician (then professor at the University of Manchester), who complained that recent measurements had much strengthened the evidence for Eddington's equation relating the masses of a proton and an electron. He feared that this confirmation of Eddington's theory (which he held to be absurd) might gain acceptance for it. His anxiety proved unjustified. A few years later I could note that a quite different set of new measurements had recently improved thirtyfold the accuracy of another prediction of Eddington's theory and this, too, was disregarded as fortuitous by the great majority of physicists. The theory has since passed into limbo— from which no conceivable future confirmation can retrieve it. In refusing to take notice of the fact that some of Velikovsky's predictions came true, scientists acted on the same lines as they did—and did rightly— in the case of Eddington's theories and in many other similar instances.[8]

This does not mean, of course, that scientists have *always* been right in so doing. I have myself suffered from the mistake of such judgements. In the same month that *The American Behavioral Scientist* came out protesting against the treatment of Velikovsky at the hand of scientific opinion, I published in *Science*[9] an account of the way my theory of the adsorption of gases on solid surfaces, that I had put forward almost half a century earlier, was disregarded by science for much of that period because its presuppositions were contrary to the current views about the nature of inter-molecular forces. I recalled how the striking evidence I had produced for the theory was shrugged aside, while flimsy observations, which have since been recognised to be misleading, were given a central

---

[8] The historical cases quoted here and in the previous four paragraphs are from my books *Science, Faith and Society* (London: Oxford University Press, 1946), *The Logic of Liberty* (London and Chicago: Routledge and University of Chicago Press, 1951), *Personal Knowledge* (London and Chicago: Routledge and University of Chicago Press, 1958).

[9] Polanyi, Michael, " The Potential Theory of Adsorption: Authority in science has its uses and its dangers ", *Science*, CXLI, 3585 (September, 1963), pp. 1010–1013.

position in contradicting my views. Yet after a while it turned out that my theory was right.

But I did not complain about this mistaken exercise of authority. Hard cases make bad laws. The kind of discipline which had gone wrong in my case was indispensable. Journals are bombarded with contributions offering fundamental discoveries in physics, chemistry, biology or medicine, most of which are nonsensical. Science cannot survive unless it can keep out such contributions and safeguard the basic soundness of its publications. This may lead to the neglect or even suppression of valuable contributions, but I think this risk is unavoidable. If it turned out that scientific discipline is keeping out a large number of important ideas, a relaxation of its severity might become necessary. But if this would lead to the intrusion of a great many bogus contributions, the situation could indeed become desperate. The pursuit of science can go on only so long as scientific judgements of plausibility are not too often badly mistaken.

The Consensual Ground of Scientific Judgement

Yet we may well wonder how the continuity enforced by current judgements of plausibility can allow the appearance of any true originality. For it certainly does allow it: science presents a panorama of surprising developments. How can such surprises be produced on effectively dogmatic grounds?

A phrase often heard in science may point towards the explanation. We speak of " surprising confirmations " of a theory. The discovery of America by Columbus was a surprising confirmation of the earth's sphericity; the discovery of electron diffraction was a surprising confirmation of de Broglie's wave-theory of matter; the discoveries of genetics brought a surprising confirmation of the Mendelian principles of heredity. We have here the paradigm of all progress in science: discoveries are made by pursuing unsuspected possibilities suggested by existing knowledge. And this is how science retains its identity through a sequence of successive revolutions.

This achievement is grounded on a tacit dimension of science, which science shares with all empirical observations. The sight of a solid object before me indicates that it has both another side and a hidden interior, which I could explore. The sight of another person indicates unlimited hidden workings of his mind and body. Perception has this inexhaustible profundity because what we perceive is *an aspect of reality*, and aspects of reality are clues to yet boundless undisclosed and perhaps as yet unthinkable experiences. This is what the existing body of scientific thought offers to the productive scientist: he sees in it an aspect of reality which as such is an inexhaustible source of new and promising problems. And his work bears this out; science continues to be fruitful, because it offers an insight into the nature of reality.

This view of science merely recognises something all scientists actually believe. For they must believe that science offers us an aspect of reality and may therefore manifest its truth inexhaustibly and often surprisingly in the future. Only in this belief can the scientist conceive problems, pursue inquiries, claim discoveries; this belief is the ground on which he teaches his students and exercises his authority over the public. And it is by transmitting this belief to succeeding generations that scientists grant their pupils independent grounds from which to start on their own discoveries and innovations—sometimes in opposition to their own teachers. This belief both justifies the discipline of scientific soundness and safeguards the freedom of scientific originality.

Admittedly, this constitution of science can work only so long as scientists have similar conceptions of the nature of things. Indeed, reasoned discussion breaks down in science between two opinions based on different foundations. Neither side can then produce an argument which the other can interpret in his own terms, as has happened in the course of a number of important scientific polemics. A famous case was the violent controversy over the question whether fermentation is caused by living cells or by inanimate catalysts. It went on for more than half a century, involving such great chemists as Woehler and Liebig on the side of " inanimate catalysts " and including Pasteur on the side of " living germ cells ". Effective argument being impossible, it was often displaced by ridicule and scorn for the opponent's views. The controversy was resolved only after the contestants had died—through the discovery in 1897 of enzymes (inanimate bodies) pressed out of living cells; in a way both sides had been right. But I can see no guarantee that such a happy resolution of a conflict will always be possible and that it would come in time to avoid a fatal break in the rational procedure of scientific argument. The continued pursuit of science would break down, if scientists came widely to disagree about the nature of things.

Implications for Scientific Policy

This brings me back to the origins of the Velikovsky case. Basic assumptions about the nature of things will tend to lie most widely apart between persons inside and outside of science. Laymen normally accept the teachings of science not because they share its conception of reality, but because they submit to the authority of science. Hence, if they ever venture seriously to dissent from scientific opinion, a regular argument may not prove feasible. It will almost certainly prove impracticable when the question at issue is whether a certain set of evidence is to be taken seriously or not. There may be nothing strange to the layman in the suggestion that the average periods of pregnancy of various animals are integer multiples of the number π, but he will only drive the scientist to despair if he challenges him to show why this is absurd. So he will be confronted with the scientist's blunt, unreasoning judgement, which rejects at a glance a set of data that seem convincing to the layman. He will demand in vain that the evidence should at least be properly examined, and will not understand why the scientist, who prides himself on welcoming any novel idea with an open mind and on holding his own scientific theories only tentatively, sharply refuses his request.

Such conflicts between science and the general public may imperil science. It is generally supposed that science will always be protected from destructive lay interference on account of its economic benefits; but this is not so. The Soviet Government adopted the theories of Lysenko and gravely hampered all branches of biological research for 30 years, overlooking altogether the damage to its agriculture. The ruling party believed in fact that it was improving the cultivation of grain through Lysenko's use of the hereditary transmission of acquired characters, an operation which scientific genetics declared impossible. Far from preventing the attack on science, the economic motive reinforced it—and this may frequently be the case. The great fallacies that have misled mankind for centuries were mostly practical.

Hence, to defend science against lay rebellions on the grounds of its technical achievements may be precarious. To pretend that science is open-minded, when it is not, may prove equally perilous. But to declare that the purpose of science is to understand nature may seem old fashioned and ineffectual. And to confess further how greatly such explanations of nature rely on vague and undemonstrable conceptions of reality may sound positively scandalous. But since all this is in fact true, might it not prove safest to say so?

In any case, it is only on these grounds that the rules governing scientific life can be understood. They alone can explain how an immense number of independent scientists, largely unknown to each other, cooperate step by step, holding the same indefinable assumptions and submitting to the same severe unwritten standards. Let us see how this works.

For a scientific community, comprising great numbers, to function, there must exist a large area of hidden and yet accessible truths, far exceeding the capacity of one man to fathom; there must be work for thousands. Each scientist starts then by sensing a point of deepening coherence, and continues by feeling his way towards such coherence. His questing imagination, guided by intuition, forges ahead until he has achieved success or admitted failure. The clues supporting his surmises are largely unspecifiable, his feeling of their potentialities hardly definable. Scientific research is one continued act of tacit integration—like making out an obscure sight, or being engaged in painting a picture, or in writing a poem. It is rare, therefore, for two scientists to contribute to one inquiry on equal terms or for one scientist to be engaged in more than one problem at a time. But it is not rare for two or more scientists to make the same discovery independently—because different scientists can actualise only the same available potentialities and they can indeed be relied on fully to exploit such chances.

This is why the initiative to scientific inquiry and its pursuit must be left to the free decision of the individual scientist; the scientist must be granted independence because only his personal vision can achieve essential progress in science. Inquiries can be conducted as surveys according to plan, but these will never add up to new ideas.

Independence will safeguard originality, which is the essence of progress in science. But there is another requirement, which must be sustained, on the contrary, by the authority of scientific opinion over scientists. We have seen scientific opinion watching that unreliable contributions should not be disseminated among scientists. But to form part of science a statement must not only be *true*, but also *interesting* and, more particularly, be interesting to science. Reliability—or exactitude—is only one factor contributing scientific interest; it is not enough. Two further important factors enter into the assessment of scientific values. One is the way a new fact enters into the systematic structure of science, correcting or expanding this structure. The other factor is independent both of the reliability and the systematic interest of a scientific discovery, for it lies in its subject matter as it was known originally, before it was investigated by science. It consists in the intrinsic *pre*-scientific interest of the object studied by science.

The scientific interest—or the scientific value—of a contribution to science is thus jointly formed by three factors: its *exactitude*, its *systematic importance*, and the *intrinsic interest of its subject matter*. The proportion in which these factors enter into scientific value varies greatly over the different domains of science; deficiency in one factor may be balanced by greater excellence in another. The highest degree of exactitude and widest range of systematisation are found in mathematical physics, and this compensates for the lesser intrinsic interest of its inanimate subject. At the other end of the sciences, we have domains like zoology and botany which lack exactitude and have no systematic structure comparable in range and beauty to that of physics, but which make up for this deficiency by the

196

far greater intrinsic interest of living things compared with inanimate matter.

A scientist engaged in research must have a keen sense of scientific value. He must be attracted by problems promising a result of scientific value and must be capable of feeling his way towards it. A great discovery may depend on realising the importance of some fact one has hit upon. Slowness in abandoning a line of inquiry which would bring only unimportant results is part of the same weakness, and most scientists who have conducted a research school have known such mortifying mistakes; they do indeed mark a deficiency in one's scientific ability.

The assessment of scientific value is also the principal standard by which the institutional structure of the scientific community is determined. Funds and appointments serving scientific research must be distributed in a way that promises the highest total increment to science. Authority at influential centres must be given to scientists distinguished by their exceptional capacity for advancing science, and rewards must be distributed at these centres in such a way as will encourage the greatest total advancement of science. Every decision of this kind requires comparisons of scientific value; it can be made in a rational manner only if there exist true standards for comparing the value of contributions to science all along the range of science from astronomy to medicine.

This does not mean that one has to compare the scientific value of one *entire branch of science* with another. It requires only that we be able to compare the value of *scientific increments* achieved in the various branches of science at similar costs of effort and money. The marginal principle of economics offers the conceptual model for this: we must try to keep equal the marginal yield, in terms of scientific value, all along the advancing borders of the sciences.

But how can anybody compare the scientific value of discoveries (and expected discoveries) in, say, astronomy with those in medicine? Nobody can, but nobody needs to. All that is required is that we compare these values in closely neighbouring fields of science. Judgements extending over neighbourhoods will overlap and form a chain spanning the entire range of sciences. This principle—*the principle of overlapping neighbourhoods*—here fulfils the functions which a capital market performs in comparing the profitability of competing enterprises among the thousand branches of an economic system.

But the two great principles of scientific growth, the granting of independence to mature scientists and the imposition of scientific values on their performances, leave open two important questions. The first question is, how can the initiatives of scientists independently choosing their problems be coordinated? The second, how can a true scientific opinion be formed, that will subject scientists to proper scientific rigour, instruct them in scientific values and cause their merits to be rightly assessed and rewarded? All these aims are achieved by two twin principles; namely, *self-coordination by mutual adjustment* and *discipline under mutual authority*.

Of *self-coordination by mutual adjustment* I have written often in the past 20 years.[10] Each scientist sets himself a problem and pursues it with a view to the results already achieved by all other scientists, who had likewise set themselves problems and pursued them with a view to the results achieved by others before. Such self-coordination represents practically the highest possible efficiency in the use of scientific talent and material resources, provided only that the professional opportunities for research and the money for its pursuit are rationally distributed.

This brings us to the principle of *mutual authority*. It consists in the fact that scientists keep watch over each other; each scientist is both subject to criticism by others and is encouraged by their appreciation. This is how scientific opinion is formed, both enforcing scientific standards and regulating the distribution of professional opportunities and research grants. Naturally, only fellow scientists working in closely related fields are competent to exercise authority over each other; but their restricted fields form chains of overlapping neighbourhoods extending over the entire range of sciences.[11]

Thus an *indirect consensus* is formed between scientists so far apart that they could not understand more than a small part of each other's subjects. It is enough that the standards of plausibility and worthwhileness be equal around every single point for this will keep them equal over all the sciences. Scientists from the most distant branches of science will rely then on each other's results and will blindly support each other against any laymen seriously challenging a scientist's professional authority.

This is the way the scientific community is organised. These are the grounds on which science rests. This the way in which discoveries are made. Science is governed by common beliefs, by values and practices

[10] The theory of mutual adjustment in society was first fully stated in my essay "The Growth of Thought in Society", *Economica*, VIII (New Series), 32 (November, 1941), and developed further in the books mentioned in footnote 8 above, as well as in my latest book, *The Tacit Dimension* (New York: Doubleday, 1966). For a summary, see Polanyi, Michael, "The Republic of Science: Its Political and Economic Theory", *Minerva*, I, 1 (Autumn, 1962), pp. 54–73.

[11] Mutual authority, based on overlapping competence, will also apply to other cultural fields and, indeed, to a wide range of consensual activities of which the participants know only a small fragment. It is a way by which resources can be rationally distributed between rival purposes that cannot be valued in terms of money. All cases of public expenditure serving collective interests are of this kind. This is, I believe, how the claims of a thousand government departments can be reasonably adjudicated, though no single person can know intimately more than a tiny fraction of them.

transmitted to succeeding generations. Each new independent member of the scientific community adheres to this tradition, assuming at the same time the responsibility shared by all members for re-interpreting the tradition and, possibly, revolutionising its teachings.

The opportunities for discovery offered by nature to the human mind are not of our making. We have developed a body of thought capable of exploiting these opportunities and have organised a body of men for this task. The laws of this community are determined by the nature of its task. But the task itself is indeterminate: it merely demands that we advance into the unknown. And even this done, the advances are not known to any single man, for no man can know more than a tiny fragment of science.

The chances offered by nature to our minds have evoked a response over which we have little control, and this is the growth of science in society. Should we rejoice to be servants of such great transcendent powers? Perhaps we are becoming—and even ought to become in other ways—a society of explorers dedicated to the pursuit of aims unknown to us. Perhaps the freedom of thought in which we take pride has such awesome implications. I cannot tell.

THE ISOLATION OF THE
SCIENTIST IN DEVELOPING COUNTRIES

Abdus Salam

Metropolis and Province in the Scientific World

Five hundred years ago—around 1470 A.D.—Saif-ud-din Salman, a young astronomer from Kandhar, working then at the celebrated observatory of Ulugh Beg at Samarkand, wrote an anguished letter to his father. In eloquent words Salman recounted the dilemmas, the heart-breaks, of an advanced research career in a poor, developing country:

"Admonish me not, my beloved father, for forsaking you thus in your old age and sojourning here at Samarkand. It's not that I covet the musk-melons and the grapes and the pomegranates of Samarkand; it's not the shade of the orchards on the banks of Zar-Afshan that keeps me here. I love my native Kandhar and its tree-lined avenues even more and I pine to return. But forgive me, my exalted father, for my passion for knowledge. In Kandhar there are no scholars, no libraries, no quadrants, no astrolabes. My star-gazing excites nothing but ridicule and scorn. My countrymen care more for the glitter of the sword than for the quill of the scholar. In my own town I am a sad, a pathetic misfit.

"It is true, my respected father, so far from home men do not rise from their seats to pay me homage when I ride into the bazaar. But some day soon all Samarkand will rise in respect when your son will emulate Biruni and Tusi in learning and you too will feel proud."

Saif-ud-din Salman never did attain the greatness of his masters, Biruni and Tusi, in astronomy. But this cry from his heart has an aptness for our present times. For Samarkand of 1470 read Berkeley or Cambridge; for quadrants read high-energy accelerators; for Kandhar read Delhi or Lahore and we have the situation of advanced scientific research and its dilemmas in the developing world of today as seen by those who feel in themselves that they could, given the opportunity, make a fundamental contribution to knowledge.

But there is one profound change from 1470. Whereas the emirate of Kandhar did not have a conscious policy for the development of science and technology—it boasted of no ministers for science, it had no councils for scientific research—the present-day governments of most developing countries would like to foster, if they could, scientific research, even advanced scientific research. Unfortunately, research is costly. Most countries do not yet feel that it carries a high priority among competing claims for their resources. Not even indigenous *applied* research can

This article from *Minerva*, IV, 4 (Summer, 1966), pp. 461–465.

command priority over straightforward projects for development. The feeling among administrators—perhaps rightly—is that it is by and large cheaper and perhaps more reliable to buy applied science on the world market. The resultant picture, so far as advanced research is concerned, remains in practice almost as bleak as at Kandhar.

Why Advanced Research Lags in Underdeveloped Countries

First and foremost among the factors that affect advanced scientific research is the supply of towering individuals, the tribal leaders, around whom great institutes are built. These are perhaps 2–3 per cent. of all men who are trained for research. What is being done in the under-developed world to ensure their supply? Most developing countries are doing practically nothing. Quite the contrary, with all the obstacles and hazards which beset a poor society, it is almost miraculous that any talent at all is saved for science. These hazards are, first, the very poor quality of education; second, the higher or administrative grades of the civil service—in India, the Indian administration service, and in Pakistan, its analogue, the civil service of Pakistan—which skim off the very top of the sub-continent's intellect; third, the poor chances for a promising young research student to learn to do research as an apprentice to a master scientist. The greatest obstacle of all lies in the very low probability of having the opportunity to work with the few men—in the case of India and Pakistan, the Siddiquis, the Usmanis, the Menons, the Sarabhais, the Seshachars—at the few centres of excellence, who appreciate at all the demands of a research career and who run laboratories which are reasonably well equipped. There are just too few scientists who retain the creativity of which they gave promise when young and there are therefore too few to train younger scientists through a fruitful master-apprentice relationship. It remains a sad fact that, though India and Pakistan may have built specialised institutes outside the university system where advanced research is carried out, by and large their vast university systems remain weak, static and uninspired. It is not part of their tradition to make a place for advanced research or even for research at all. The colleges which provide a very large proportion of under-graduate education in India and Pakistan have grown up in a tradition of concentrating such resources as they have on the instruction and moral formation of undergraduates. I shall always remember my first interview with the head of the premier college in Pakistan, which I joined after a spell of theoretical work in high-energy physics at Cambridge and Prince-ton. My chief said: "We all want research men here, but never forget we are looking more for good, honest teachers and good honest college men. This college has proud traditions to uphold. We must all help.

Now for any spare time you may have after your teaching duties, I can offer you a choice of three college jobs: you can take on wardenship of the college hostel; or be chief treasurer of its accounts; or if you like, become president of its football club ". As it was, I was fortunate to get the football club.

Admittedly, this was 12 years ago. I should be ungrateful if I did not mention that this same college today is contending with the Atomic Energy Commission of Pakistan for the control of a high-tension laboratory with a 2·5 Mev Cockcroft-Walton set. This is a measure of the change brought about by the heroic efforts of the Pakistan Government since 1958. Things have changed. Nonetheless the situation of advanced research in underdeveloped countries still remains greatly in need of help.

In a number of fields, advanced scientific research in developing countries is beginning to reach the stage of maturity in which first-rate work can be done. Indigenous resources are being skilfully employed but there is still a desperate need for international help. The truth is that, irrespective of a man's talent, there are in science, as in other spheres, the classes of haves and have-nots; those who enjoy the physical facilities and the personal stimulus for the furtherance of their work, and those who do not, depending on which part of the world they live in. This distinction must go. The time has come when the international community of scientists should begin to recognise its direct moral responsibility, its direct involvement, its direct participation in advanced science in developing countries, not only through helping to organise institutions but by providing the personal face-to-face stimulation necessary for the first-rate individual working in these countries.

In advanced scientific research, it is the personal element that counts much more than the institutional. If, through meaningful international action, allied with national action, we could build the morale of the active research worker and persuade him not to make himself an exile, we shall have won a real battle for the establishment of a creative scientific life in the developing countries.

Breaking the Barrier of Isolation

As an example of what is needed, I shall take the science with which I am personally associated. Theoretical physics happens to be one of the few scientific disciplines which, together with mathematics, is ideally suited to development in a developing country. The reason is that no costly equipment is involved. It is inevitably one of the first sciences to be developed at the highest possible level; this was the case in Japan, in India, in Pakistan, in Brazil, in Lebanon, in Turkey, in Korea, in Argentina. Gifted men from these countries work in advanced centres in the West or the Soviet Union. They then go back to build their own indigenous schools. In the past, when these men went back to the universities in their home countries, they were perhaps completely alone; the groups of which they formed a part were too small to form a critical mass; there were no good libraries, there was no communication with groups abroad. There was no criticism of what they were doing; new ideas reached them too slowly; their work fell back within the grooves of what they were doing before they left the stimulating environments of the institutions at which they had studied in the West or the Soviet Union. These men were isolated, and isolation in theoretical physics—as in most fields of intellectual work—is death. This was the pattern when I became associated with Lahore University; this is still the pattern in Chile, in Argentina, in Korea.

In India and Pakistan we have been more fortunate than most other underdeveloped countries in the last decade. A number of specialised institutes have grown up for advanced work in theoretical physics—the Tata Institute at Bombay, the Institute of Mathematical Sciences at Madras and the Atomic Energy Centres at Lahore and Dacca—where a fair concentration of good men exists. But this is not enough. These institutes are still small oases. They are too small for the fertilisation of the area around them. They are also in continuous danger of being dried up because the area around them is too arid and they still do not have vigorous contacts with the world community. Tata and Madras have partly solved their problem; they have funds to invite visitors—they have fewer funds to send Indian physicists abroad, mainly because of the serious shortage of foreign exchange.

It was with this type of problem in mind that the idea of setting up an International Centre for Theoretical Physics [1] was mooted in 1960. The idea was to establish a truly international centre, run by the United Nations family of organisations, for advanced research in theoretical physics. It was planned with two objectives in view: first, to bring physicists from the East and the West together; second, and even more important, to provide extremely liberal facilities for senior active physicists from developing countries.

The International Centre tries to deal with the problem of isolation in a number of ways. We have ordinary fellowships which are given mostly to those from developing countries. In addition, the International Centre has instituted an associateship scheme. A number of carefully selected senior active physicists from developing countries are given the privilege of coming for a period of one to four months every year to the centre with no prior formalities other than a letter to the director announcing

[1] Cf., *Minerva*, III, 4 (Summer, 1965), pp. 533-536.

their arrival. The centre pays for their transportation and their maintenance. The aim is to have eventually, at any one time, a group of about 50 senior active physicists from developing countries who possess this privilege.

Looking back on my own period of work in Lahore, as I said, I felt terribly isolated. If at that time someone had said to me, we shall give you the opportunity every year to travel to an active centre in Europe or the United States for three months of your vacation to work with your peers; would you then be happy to stay the remaining nine months at Lahore, I would have said yes. No one made the offer. I felt then and I feel now that this is one way of halting the brain drain, of keeping active men happy and contented within their own countries. They must be kept there to build for the future, but their scientific integrity must also be preserved. By providing them with this guaranteed opportunity for remaining in contact with their peers, we believe we are making a contribution to solving the problem of isolation.

Ideally the associateship scheme should be wide enough to cover nearly every active physicist in developing countries. It should be well publicised; every first-rate research worker should know and feel confident that he could almost, as it were, demand its privileges if he were living in a developing country. Unfortunately, the International Centre at Trieste does not possess funds to do this. Yet the scheme is not very costly. Since it pays no salaries—only the fare and a *per diem* allowance—it costs us something like $100,000. Since the associateship scheme seems thus far to be the most fruitful of all the available ways for breaking the isolation which kills the creativity of creative scientists, it should be extended.

Universities and institutions with the wealth and scientific eminence of Princeton, Harvard, Cambridge, All Souls, Rockefeller University, New York State University, the Imperial College in London and others should seriously consider the establishment of their own associateship schemes. It ought to be considered not only for theoretical physics but for other subjects too. Rockefeller University, for example, might extend the privilege of giving its freedom not only to a scientist of the distinction of Professor Seshchar, but also to other active micro-biologists in most developing countries. The European Organisation for Nuclear Research at Geneva has already started a scheme similar to our own, which, I believe, covers both experimental and theoretical physics. It is designed of course only for less-developed countries within Europe (Greece and Spain).

If every active, first-rate worker in the developing countries could be covered, we would go very far towards the removal of one of the curses of being a scientist in a developing land.

INDEX

205